RECENT ADVANCES IN VIBRATIONS ANALYSIS

Edited by **Natalie Baddour**

INTECHOPEN.COM

Recent Advances in Vibrations Analysis
Edited by Natalie Baddourr

CBS, Edition 2016

Published by InTech
Janeza Trdine 9, 51000 Rijeka, Croatia

Publishing Process Manager Dragana Manestar

Technical Editor Teodora Smiljanic

Cover Designer InTech Design Team

Image Copyright Eskemar, 2011. Used under license from Shutterstock.com

Additional hard copies can be obtained from orders@intechopen.com

Recent Advances in Vibrations Analysis, Edited by Natalie Baddour
p. cm.
ISBN 978-953-307-696-6

INTECH
open science | open minds

Contents

Preface

This book covers recent advances in modern vibrations analysis, from analytical methods to applications of vibrations analysis to condition monitoring.

The book opens with a section on recent advances in analytical methods. Dr. Kang Bongsu contributed a chapter that presents an alternative technique for the free and forced vibration analysis of one-dimensional distributed parameter systems. This approach is based on the idea of superimposing the amplitudes of the constituent travelling waves, rather than the traditional approach of normal mode expansion that relies on the apriori calculation of eigensolutions or assumed normal modes.

In the second chapter, Juan Carlos Jáuregui presents an application of phase space to the identification of nonlinearities and transients. In this interesting approach, a phase diagram is represented as a three-dimensional plot which can then be used for frequency and dynamic identification of a system. The application of this approach to nonlinear mechanical systems such as gears, bearings and friction is also included in the chapter.

The next chapter presents an analytical solution for a nano-plate with Levy boundary conditions. The free vibration analysis is based on a first order shear deformation theory which includes the small scale effect. The governing equations of motion, reformulated as two new equations called the edge-zone and interior equations, are based on the nonlocal constitutive equations of Eringen.

A. Shahrjerdi and F. Mustapha co-authored the fourth chapter, which discusses second-order shear deformation theory applied to a plate with simply supported boundary conditions. The material properties of the plate are graded in the thickness direction by a power law distribution and the equations of motion are derived via the energy method and then solved by applying Navier's method. It is interesting to note that the authors demonstrate that the results of the second-order theory are very close to those reported in the literature using a third-order theory.

The next section of the book deals with the application of vibrations analysis to the condition monitoring and maintenance of various machines. The first chapter in this section deals with the maintenance of reducers that have unbalanced loads. The most commonly used maintenance approaches for reducers are oil analysis (via laboratory chemical analysis) and separately, vibrations analysis. In this chapter, a novel way of combining the two approaches for more accurate results is presented.

The second chapter in this section presents an alternative method for detecting failures in transformers via the analysis of the vibrations produced inside the transformer under operation. Normally, the transformer produces vibrations in the windings and the core, and these vibrations vary according to operating conditions. However, in the presence of mechanical failure, the vibration patterns are different from those produced by normal conditions. This idea is used as the basis for a failure detection mechanism, with the promise that this approach makes it possible to design an on-line real-time diagnosis system.

The final chapter in this section describes an interesting method for monitoring the thermal snap of satellites, an effect which has been known to cause attitude disturbance in Low Earth Orbit satellites. The difficulty with these types of thermally induced vibrations is that they are very slow and cannot be monitored via a traditional sensor-driven approach. This chapter thus describes a novel approach to this problem via an onboard monitoring camera. Images taken in space and the image processing of these images are explained.

The third and final section of the book deals with the modelling and analysis of various complex mechanical systems. In particular, the first chapter of this section deals with the vibrations analysis of several mechanical systems possessing complex design and geometry. Specific systems considered include a fatigue test rig for aviation gear boxes, a gas turbine blade and finally an annular membrane resting on an elastic foundation of a Winkler type.

The next chapter in this section considers the free vibration of eccentric building systems. In particular, the coupled torsional-translational vibrations of both symmetric and eccentric one-storey building systems subjected to ground excitation are modelled and then analysed.

In the subsequent chapter, the structural modelling of beams as part of the hydroelastic analysis of large container ships is presented. In developing these models, it is important to appropriately account for the contribution of transverse bulkheads to hull stiffness and the behaviour of the relatively short engine room structure. The application of this approach to the hydroelastic analysis of a very large container ship is then illustrated.

The final chapter deals with the use of the stochastic finite element method for vibrations analysis. Although the finite element method analysis of complicated structures has become generally accepted, regarding the given factors as known constants does not always correspond to the reality that material properties, geometry parameters and applied loads of the structure are often modelled as stochastic. Thus, how to incorporate the stochastic nature of these parameters into a finite element model is shown.

I would like to express warm thanks to all the contributors, in particular for their efforts to ensure that difficult material is made accessible to wider audience.

Dr. Natalie Baddour
Department of Mechanical Engineering
University of Ottawa
Canada

Part 1

Analytical Methods

1

Exact Transfer Function Analysis of Distributed Parameter Systems by Wave Propagation Techniques

Bongsu Kang
Indiana University – Purdue University Fort Wayne
USA

1. Introduction

The vibrations of elastic structures such as strings, beams, and plates can be described in terms of waves traveling in waveguides (Cremer et al., 1973; Graff, 1975; Fahy, 1987). While the subject of wave propagation has been extensively studied in the fields of acoustics in fluids and solids rather than vibrations of elastic structures, wave analysis techniques have been employed to reveal physical characteristics associated with structural vibrations of elastic media (Argento & Scott, 1995; Kang & Tan, 1998). One of the advantages of the wave analysis technique, when applied to the structural vibration analysis, is its compact and systematic approach to analyze complex structures with discontinuities (Mace, 1984; Yong & Lin, 1989; Kang et al., 2003; Mei & Mace, 2005). Applying the concept of wave reflection and transmission, Mace (1984) obtained the frequency equations of Euler-Bernoulli beams including waves of both propagating and near-field types. By the phase-closure principle, also referred to as the wave-train closure principle (Cremer et al., 1973), Mead (1994) determined natural frequencies of Euler-Bernoulli beams. This principle states that if the phase difference between incident and reflected waves is an integer multiple of 2π, then the waves propagate at a natural frequency and their motions constitute a vibration mode. Based on the same principle, Kang (2007) presented a systematic approach to the free and forced vibration analysis of multi-span beams.

The classical method, known as the normal mode or eigenfunction expansion, of solving the forced vibration problem of a distributed parameter system involves expansion of the forcing function into the eigenfunctions of the associated free vibration problem. While this method is theoretically sound and powerful, the method is difficult to implement when the problem to be solved is a non-self-adjoint system typically due to complicating effects such as damping, discontinuities, or non-classical boundary conditions, for which case obtaining the exact eigensolutions is not often feasible. Although approximate eigensolutions may be used instead of exact ones, the problem still persists in the form of poorly convergent solution and/or significant error in the solution. As an alternative approach to solve forced vibration problems, Yang and Tan (1992) presented a method for evaluating exact closed-form transfer functions for a class of one-dimensional distributed parameter systems. Applying the energy functionals of constrained and combined damped systems, Yang (1996a, 1996b) presented a method to obtain a closed-form transient response solution in eigenfunction series for a distributed damped system.

The dynamic displacement of any point in an elastic waveguide can be determined by superimposing the amplitudes of the constituent waves traveling along the waveguide, which is a basis of wave propagation. Based on this simple fact, an alternative technique for the free and forced vibration analysis of one-dimensional distributed parameter systems is presented. The method of normal mode expansion is often difficult to implement for nonself-adjoint systems with complicating effects such as mode couplings, non-proportional damping, discontinuities, or arbitrary boundary conditions, since the method requires eigensolutions or assumed normal modes as a priori. However, this alternative analysis technique based on the elastic wave propagation does not pose such a requirement and leads to the exact, closed-form, distributed transfer function of a distributed parameter system. The general wave solution of the equation of motion governing the dynamics of a waveguide is cast into a matrix form in terms of the constituent waves defined in the Laplace domain. The spatial amplitude variation of the traveling wave is represented by the field transfer matrix and the amplitude distortion of the traveling wave incident upon a discontinuity due to geometric or kinetic constraints is described by the local wave reflection and transmission matrices. Combining these matrices in a progressive manner along the waveguide by applying the concepts of global wave reflection and transmission matrices leads to the exact characteristic equation and corresponding mode shapes for the free response analysis and the transfer function of the system for the forced response analysis. The transient response solution for a complex system can be obtained through the Laplace inversion of the transfer function using numerical inversion algorithms. The exact frequency response solution, which includes infinite normal modes of the system, can be obtained in terms of the complex frequency response function from the transfer function. One of the main advantages of this analysis technique is its systematic formulation resulting in a recursive computational algorithm which can be implemented into highly efficient computer codes. This systematic approach also allows modular formulation which can be readily expandable to include additional discontinuities with little alteration to the existing formulation. In addition, it is also computationally advantageous that the technique always results in operating matrices of a fixed size regardless of the number of discontinuities in a waveguide. This analysis technique is applicable to any one-dimensional waveguides (strings, axial rods, torsional bars, beams, and frame structures), in particular systems with multiple point discontinuities such as viscoelastic supports, attached inertias, and geometric/material property changes. The analysis technique is demonstrated using the second order wave equation, fourth order beam equation, and sixth order curved beam equation.

2. Second order systems

The free transverse vibration of a taut string, longitudinal vibration of a thin bar, and the torsional vibration of a shaft are governed by the equation of motion in the same form, the wave equation, and thus they are mathematically analogous. Therefore, with no loss of generality, the transverse vibration of the string is taken as a representative problem for the description of the present analysis technique based on wave propagation. The equation governing the transverse motion of a uniformly damped string of span length L is

$$T\frac{\partial^2 W}{\partial X^2} + C_e\frac{\partial W}{\partial t} = m\frac{\partial^2 W}{\partial t^2} \quad (2.1)$$

where W is the transverse displacement, X the spatial variable, t the temporal variable; and T denotes the constant tension, C_e the damping coefficient, and m the mass per unit length of the string. With introduction of the following non-dimensional variables and parameters

$$w = W/L \quad x = X/L \quad \tau = t/t_0 \quad c_e = C_e L/t_0 \quad t_0 = L/c_0 \quad c_0 = \sqrt{T/m} \quad (2.2)$$

the equation of motion takes the non-dimensional form of

$$w'' + c_e\dot{w} = \ddot{w} \quad (0 \le x \le 1) \quad (2.3)$$

where the prime (′) and dot (·) denote the differentiation with respect to x and τ, respectively. Applying the Laplace transform to Eq. (2.3) yields

$$\overline{w}''(x;s) + c_e s\overline{w}(x;s) = s^2\overline{w}(x;s) \quad (2.4)$$

where s denotes the Laplace variable and zero initial conditions are assumed.

2.1 Wave solution

Denoting C as the amplitude of the wave traveling along the string, the solution of Eq. (2.4) can be assumed in wave form

$$\overline{w}(x;s) = Ce^{i\gamma x} \quad (2.5)$$

where γ is the non-dimensional wavenumber normalized against span length L. Applying the above wave solution to Eq. (2.4) gives the frequency equation of the problem

$$\gamma^2 + s^2 - c_e s = 0 \text{ or } \gamma^2 = c_e s - s^2 \quad (2.6)$$

from which the general wave solution can be found as the sum of two constituent waves

$$\overline{w}(x;s) = C^+e^{-i\gamma x} + C^-e^{i\gamma x} \quad (2.7)$$

where the coefficient C represents the amplitude of each wave component with its traveling direction indicated by the plus (+) or (–) sign. Note that γ is complex valued for nonzero c_e, hence the classification of the wave into propagating and attenuating waves does not apply to this case. Defining $f(x;s) = e^{-i\gamma x}$ as the *field transfer function* which relates the wave amplitudes by

$$C^+(x + x_0) = fC^+(x_0) \text{ or } C^-(x + x_0) = f^{-1}C^-(x_0) \quad (2.8)$$

where $f^{-1}(x;s) = e^{i\gamma x}$, the wave solution in Eq. (2.7) can be re-written as

$$\overline{w}(x;s) = fC^+ + f^{-1}C^- \quad (2.9)$$

2.2 Wave reflection and transmission

When a wave traveling along a string is incident upon a discontinuity such as an elastic support, geometric/material property change, or boundary, it is reflected and transmitted at different rates depending on the properties of the discontinuity. The rates of wave reflection and transmission can be determined in terms of *wave reflection* and *transmission coefficients*. For example, consider an infinitely long string constrained at a local coordinate $\xi = 0$ as shown in Fig. 1, where the constraint is a point support consisting of an attached mass (M_c), a transverse spring (K_c), and a viscous damper (C_c).

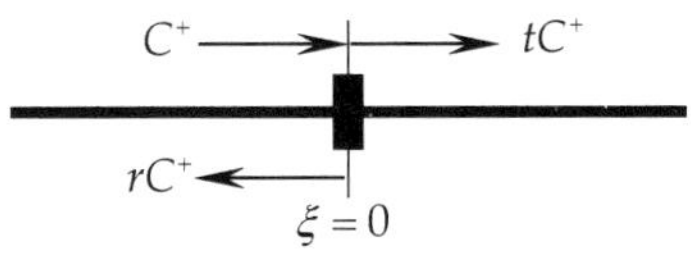

Fig. 1. Wave reflection and transmission at a discontinuity.

When a positive-traveling wave C^+ is incident upon the support, it gives rise to a reflected wave rC^+ and a transmitted wave tC^+, where r and t represent the wave reflection and transmission coefficients, respectively. The resulting transverse displacements at the left and right of the constraint are

$$\bar{w}_l = C^+ + rC^+ \quad \bar{w}_r = tC^+ \tag{2.10}$$

Since $\bar{w}_l = \bar{w}_r$ at ξ=0, one can find

$$1 + r = t \tag{2.11}$$

In addition, the kinetic equilibrium condition at ξ=0 states that

$$\bar{w}_r' - \bar{w}_l' = \varphi \bar{w}_r \tag{2.12}$$

where φ represents the kinetic properties of the constraint as

$$\varphi(s) = k_c + c_c s + m_c s^2 \tag{2.13}$$

Note that k_c, c_c, and m_c in Eq. (2.13) are the non-dimensional spring constant, damping coefficient, and the attached mass, respectively, defined by

$$k_c = K_c L/T \quad c_c = C_c c_0/T \quad m_c = M_c c_0^2/TL \tag{2.14}$$

Equation (2.12) leads to

$$\gamma(r + t - 1) = i\varphi t \tag{2.15}$$

Combining Eqs. (2.11) and (2.15), the wave reflection and transmission coefficients can be found as

$$r = \frac{i\varphi}{2\gamma - i\varphi} \quad \text{and} \quad t = \frac{2\gamma}{2\gamma - i\varphi} \tag{2.16}$$

When a wave is incident upon a boundary at ξ=0, it is only reflected, therefore

$$\bar{w}_l' = -\varphi \bar{w}_l \tag{2.17}$$

which leads to

$$r = \frac{\gamma + i\varphi}{\gamma - i\varphi} \tag{2.18}$$

In the limiting case where $\varphi=\infty$, it can be seen that $r=-1$ which is the wave reflection coefficient for the classical fixed boundary. When the wave is incident upon a series of discontinuities along its traveling path, it is more computationally efficient to employ the concepts of *global wave reflection and transmission coefficients*, in particular when the free or forced vibration analysis of a multi-span string is sought. These coefficients relate the amplitudes of incoming and outgoing waves at a discontinuity. Consider wave motion in a multi-span string as illustrated in Fig. 2. Define R_{ir} as the global wave reflection coefficient which relates the amplitudes of negative- and positive-traveling waves on the right side of discontinuity i such that

$$C_{ir}^- = R_{ir} C_{ir}^+ \tag{2.19}$$

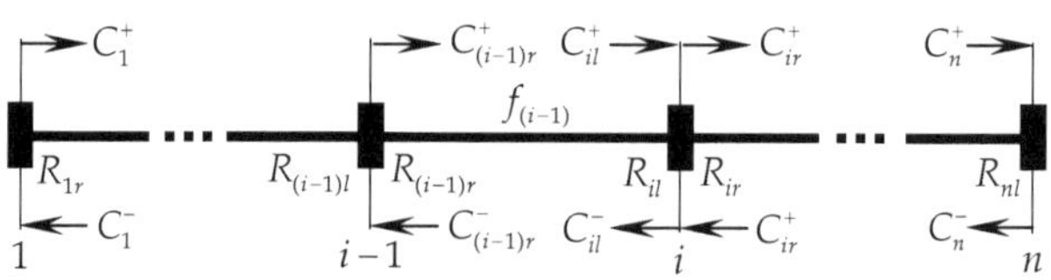

Fig. 2. Waves traveling along a multi-span string.

Since $C_{ir}^- = f_i R_{(i+1)l} C_{ir}^+$, one can find R_{ir} in terms of the global wave reflection coefficient on the left side of discontinuity i+1; i.e.,

$$R_{ir} = f_i^2 R_{(i+1)l} \tag{2.20}$$

In addition, by combining the following wave equations at discontinuity i

$$C_{ir}^+ = t_i C_{il}^+ + r_i C_{ir}^- \text{ and } C_{il}^- = t_i C_{ir}^- + r_i C_{il}^+ \tag{2.21-22}$$

the relationship between the global wave reflection coefficients on the left and right sides of discontinuity i can be found as

$$R_{il} = r_i + t_i^2 (R_{ir}^{-1} - r_i)^{-1} \tag{2.23}$$

R_{ir} and R_{il} progressively expand to include all the global wave reflection coefficients of discontinuities along the string before terminating its expansion at the boundaries where

$$C_1^+ = r_1 C_1^- \quad C_n^- = r_n C_n^+ \tag{2.24}$$

While the global wave reflection coefficient relates the amplitudes of waves traveling in the opposite direction of each other within a subspan *i*, there is a need for another coefficient which relates the amplitudes of waves traveling in the same direction in two adjacent subspans. This inter-span coefficient is particularly useful when the mode shape or forced response of a string with several or more subspans needs to be determined since it allows an intuitive and systematic formulation of the system's transfer function. Denoting this inter-span wave transfer coefficient as the *global wave transmission coefficient* T_i, define

$$C_{ir}^{+} = T_i C_{(i-1)r}^{+} \tag{2.25}$$

Rewriting Eq. (2.21) by applying $C_{il}^{+} = f_{(i-1)} C_{(i-1)r}^{+}$ and $C_{ir}^{-} = R_{ir} C_{ir}^{+}$, and then comparing it with Eq. (2.25), the global wave transmission coefficient at discontinuity *i* can be found as

$$T_i = (1 - r_i R_{ir})^{-1} t_i f_{(i-1)} \tag{2.26}$$

The global wave reflection and transmission coefficients are the key elements in determining the exact transfer function of a multi-span string as discussed in the following sections.

2.3 Free response analysis

The global reflection and transmission coefficients of waves traveling along a multi-span sting are now combined with the field transfer function to analyze the free response of a multi-span sting. With reference to Fig. 2, consider wave motion in the first span. Based on the definition of the global wave reflection coefficient, at the boundary

$$C_1^{-} = R_{1r} C_1^{+} \tag{2.27}$$

However, recalling $C_1^{+} = r_1 C_1^{-}$ from Eq. (2.24), it can be found that

$$(r_1 R_{1r} - 1) C_1^{+} = 0 \tag{2.28}$$

For nontrivial solutions,

$$F(s) = r_1 R_{1r} - 1 = 0 \tag{2.29}$$

which is the characteristic equation in terms of the Laplace variable *s* for the multi-span string with arbitrary discontinuities and boundaries. This remarkably simple expression for the characteristic equation is due to the fact that R_{1r} recursively expands to include all the effects of constraints in the remaining side of the string until its expansion terminates at the rightmost boundary which yields $R_{nl}=r_n$. This equation simply states that when the string system vibrates at one of its natural frequencies, $r_1 R_{1r}=1$. As a simple example, for a single span uniformly damped string fixed at both ends, $r_1=r_2=-1$ from Eq. (2.18) and $R_{1r}=e^{-2i\gamma}$ from Eq. (2.20). Therefore, $F(s)=e^{-2i\gamma}-1=0$ gives the natural wavenumbers $\gamma_n=n\pi$ (n=1,2,3,...), or in terms of the non-dimensional frequency $\omega_n^2 + i c_e \omega_n = (n\pi)^2$ by Eq. (2.6) with *s* replaced by $i\omega$.

The mode shapes of the multi-span string system can be found in a systematic manner by relating wave amplitudes between two adjacent subspans. Let ξ_i denote the local coordinate within span *i*. The transverse displacement at any point in span *i* can be expressed as

$$\overline{w}_i(\xi_i) = f_i(\xi_i)C_i^+ + f_i^{-1}(\xi_i)C_i^- \tag{2.30}$$

However, due to Eq. (2.19)

$$\overline{w}_i(\xi_i) = [f_i(\xi_i) + f_i^{-1}(\xi_i)R_{ir}]C_i^+ \tag{2.31}$$

Now by applying the global wave transmission coefficient defined in Eq. (2.25) to C_i^+ in the above equation, the displacement at any point in span i of the string can be expressed in terms of the wave amplitude in span i–1 as

$$\overline{w}_i(\xi_i) = [f_i(\xi_i) + f_i^{-1}(\xi_i)R_{ir}]T_iC_{i-1}^+ \tag{2.32}$$

Assume a disturbance arise in span 1; i.e., a wave originates and starts traveling from the leftmost boundary of the string. Then, by successively applying the global transmission coefficient of each discontinuity on the way up to the first span, the mode shape of span i can be found in terms of wave amplitude C_1^+; i.e.,

$$\overline{w}_i(\xi_i) = [f_i(\xi_i) + f_i^{-1}(\xi_i)R_{ir}]\prod_{j=i}^{1} T_jC_1^+ \quad C_1^+ \quad 0 \le \xi_i \le l_i \tag{2.33}$$

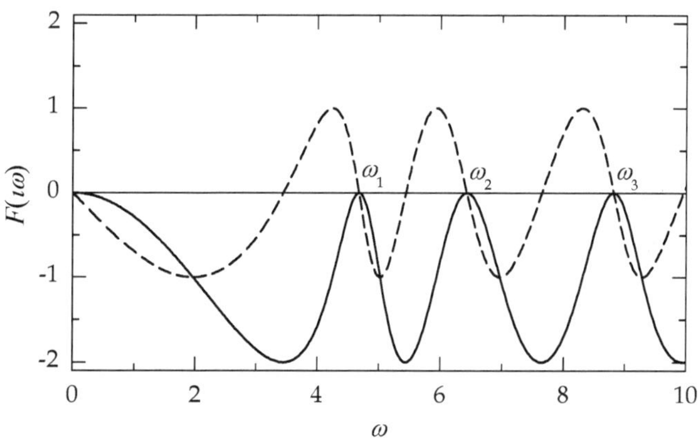

Fig. 3. Plot of the characteristic equation. The solid and dashed curves represent the real and imaginary parts, respectively.

For example, consider a fixed-fixed undamped string with three supports specified by φ_2=5+0.1s^2, φ_3=7+0.1s^2, and φ_4=4+0.1s^2 according to Eq. (2.13). l_1=0.25, l_2=0.3, l_3=0.25, and l_4=0.2 are assumed. Once the global wave reflection coefficient at each discontinuity has been determined, one can apply Eq. (2.29) to find the natural frequencies. Shown in Fig. 3 is the plot of the characteristic equation, where the first three natural frequencies are indicated. The mode shapes can be found from Eq. (2.33) in a systematic way once the global wave transmission coefficient at each discontinuity has been determined. Figure 4 shows the mode shapes for the first three modes.

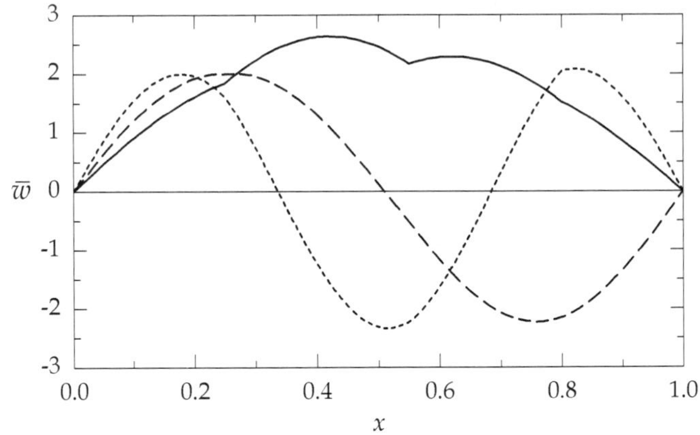

Fig. 4. First three mode shapes obtained from Eq. (2.33). The solid, dashed, and dotted curves represent the 1st , 2nd, and 3rd modes, respectively.

2.4 Transfer function analysis

Consider a multi-span string subjected to an external point load $\bar{p}(s)$, normalized against tension T, applied at x=x_0 as shown in Fig. 5. Since the waves injected by the load travel in both direction from the point of loading, a set of local coordinates $\{\xi, \xi^*\}$ is introduced such that the wave traveling toward each boundary of the string is considered positive as indicated in Fig. 5. Let $C_1^{\pm}$ and $D_1^{\pm}$ denote the injected waves that travel in the region x<x_0 and $x \geq x_0$ within span 1, respectively. The transverse displacement of span 1 of the string can be expressed as

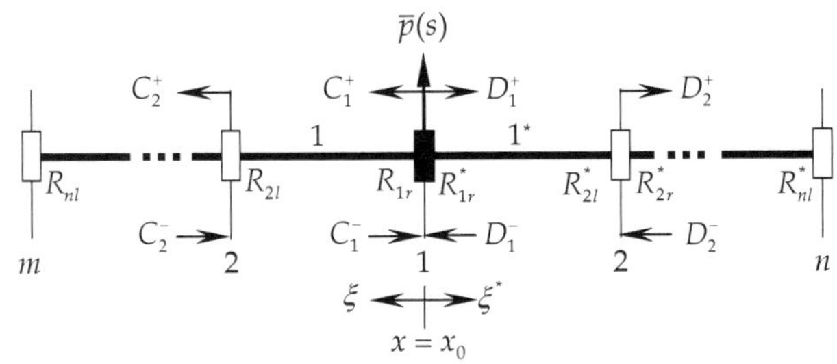

Fig. 5. Wave motion in a multi-span string due to a point load.

$$\bar{w}_1(\xi_1, x_0; s) = f_1(\xi_1; s)C_1^+ + f_1^{-1}(\xi_1; s)C_1^- \quad 0 \leq \xi_1 \leq l_1 \tag{2.34}$$

Defining v as the wave injected by the applied load, the difference in amplitudes between the incoming and outgoing waves at the loading point is

$$C_1^+ - D_1^- = v \quad D_1^+ - C_1^- = v \tag{2.35}$$

where v can be determined from the geometric and kinetic continuity conditions at ξ=0 as

$$v = \frac{\sqrt{-1}}{2\gamma}\bar{p}(s) \tag{2.36}$$

Applying the global wave reflection coefficient on each side of the loading point gives

$$C_1^- = R_{1r}C_1^+ \quad D_1^- = R_{1r}^* D_1^+ \tag{2.37}$$

where the asterisk (*) signifies that the global wave reflection coefficient is defined in the region $x \geq x_0$ to distinguish it from the one defined in the region $x < x_0$. Combining Eqs. (2.35) and (37), one can determine the ampltitude of the wave that rises at the loading point in each direction

$$C_1^+ = T_1 v \quad T_1 = (1 - R_{1r}^* R_{1r})(1 + R_{1r}^*) \tag{2.38.1}$$

$$D_1^+ = T_1^* v \quad T_1^* = (1 - R_{1r} R_{1r}^*)(1 + R_{1r}) \tag{2.38.2}$$

Note that T_1 and T_1^* and can be considered as the global wave transmission coefficients that characterize the transmissibility of the wave injected by the external force in the region $x<x_0$ and $x \geq x_0$, respectively. It is evident that and T_1 and T_1^* are different unless the string system is symmetric about the loading point. Applying the results in Eqs. (2.37) and (2.38) to Eq. (2.34), the wave motion in either side of $x=x_0$ can be found; i.e., in the region $x<x_0$

$$\bar{w}_1(\xi_1, x_0; s) = [f_1(\xi_1; s) + f_1^{-1}(\xi_1; s)R_{1r}]T_1 v \quad 0 \leq \xi_1 \leq l_1 \tag{2.39}$$

Now, in the same manner as for the mode shape analysis, since $C_{i+1}^+ = T_i C_i^+$ and $C_1^+ = T_1 v$, the wave motion in span i on either side of the loading point can be found. For the region $x<x_0$

$$\bar{w}_i(\xi_i, x_0; s) = [f_i(\xi_i; s) + f_i^{-1}(\xi_i; s)R_{ir}]\prod_{k=i}^{1} T_k v \quad 0 \leq \xi_i \leq l_i \tag{2.40}$$

Note that the ratio

$$G_i(\xi_i, x_0; s) = \bar{w}_i(\xi_i, x_0; s)/\bar{p}(s) \tag{2.41}$$

is the transfer function governing the forced response of any point in span i due to the point loading at $x=x_0$. The Laplace inversion of $G_i(\xi_i, x_0; s)$ is the Green's function of the problem. Thefore, denoting L^{-1} as the inverse Laplace transform operator, the forced response at any point within any subspan of the multi-span string can be determined from the following convolution integral; e.g., for span i in the region $x<x_0$

$$w_i(\xi_i, x_0; \tau) = \int_0^{\tau} G_i(\xi_i, x_0; \tau - \bar{\tau})p(\tau)d\bar{\tau} \quad (i = 1, 2, \cdots, m) \tag{2.42.1}$$

$$G_i(\xi_i, x_0; \tau) = L^{-1}[\frac{\sqrt{-1}}{2\gamma}[f_i(\xi_i; s) + f_i^{-1}(\xi_i; s)R_{ir}]\prod_{k=i}^{1} T_1] \quad 0 \leq \xi_i \leq l_i \tag{2.42.2}$$

The exact Laplace inversion of $G_i(\xi_i, x_0; s)$ in close form is not feasible in general, in particular for multi-span string systems. One may have to resort to the numerical inversion of Laplace transforms. It is found that the algorithm known as the fixed Talbot method (Abate & Valko, 2004), which is based on the contour of the Bromwich inversion integral, appears to perform the numerical Laplace inversion in Eq. (2.42) with a satisfactory accuracy and reasonable computation time. In this method, the accuracy of the results depends only on the number of precision decimal digits (denoted with M in the algorithm) carried out during the inversion. It is found that for well-damped second- and higher order distributed parameter systems, M=32 and for lightly damped or undamped systems, M=64 gives acceptable results. To demonstrate the effectiveness of the present analysis approach, consider the unit impulse response of a single span undamped string, in particular the response near the point of loading immediately after the loading, which is well known for its deficiency in numerical convergence. The response solution by the method of normal mode expansion is

$$w(x,x_0;\tau)=\sum_{n=1}^{N}\frac{2\sin n\pi x_0}{n\pi}\sin n\pi x\sin n\pi\tau \tag{2.43}$$

The corresponding transfer function from Eq. (2.40) is

$$\overline{w}(x,x_0;s)=\frac{1}{2s(e^{2s}-1)}\begin{cases} e^{-sx}(e^{2s(1-x_0)}-1)(e^{2sx}-e^{2sx_0}) & \text{for } 0\le x\le x_0 \\ e^{-sx}(e^{2sx_0}-1)(e^{2sx}+e^{2s(1-x_0)}) & \text{for } x_0< x\le 1 \end{cases} \tag{2.44}$$

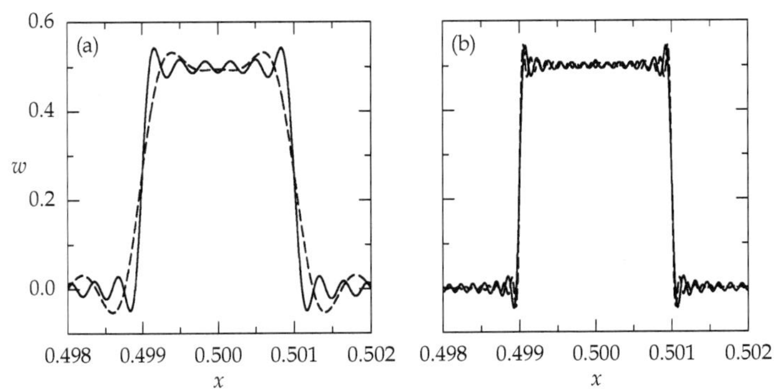

Fig. 6. Unit impuse response of a single span string near the point of loading at τ=0.001: (a) M=32 and N=2,500; (b) M=64 and N=20,000. The solid and dashed curves represent the solutions from Eqs. (2.43) and (2.44), respectively.

Shown in Fig. 6 is the comparison of the response solutions given in Eqs. (2.43) and (2.44) when τ=0.001 near x=x_0=0.5. N=2,500 and N=20,000 are used for the evaluation of the series solution, while M=32 and M=64 are used for the numerical Laplace inversion of the transfer function in Eq. (2.44). It can be seen that the series solution in Eq. (2.43) fails to accurately represent the actual impulse response behavior with N=2,500. This is expected for the series solution since it would take a large number of harmonic terms (N≥10,000 for this example) to represent such a sharp spike due to the impulse. It can be seen that the result with M=32 reasonably represents the actual behavior, and the result with M=64 is almost comparable to the series solution with N=20,000. However, if one tries to obtain the response at a time very close to the moment of impact, the numerical Laplace inversion becomes extremely strenuous or beyond the machine precision of the computing machine. This is because the expected response would consist of waves that have unrealistically short wavelengths. This is not a unique problem for the present wave approach since the same problem would manifest itself in the series solution given in Eq. (2.43), requiring an impractically large number of harmonics terms for a convergent solution.

If $p(\tau)=p_0 e^{i\omega\tau}$; i.e., a harmonic forcing function, the steady-state response of the problem can be readily found in terms of the complex frequency function defined as

$$H(\xi_i, x_0; \omega) = G(\xi_i, x_0; s)|_{s=i\omega} \tag{2.45}$$

Therefore the frequency response at any point within any subspan of the string can be obtained by; e.g., for span i in the region $x<x_0$

$$w_i(\xi_i, x_0; \tau) = |H_i(\xi_i, x_0; \omega)| p_0 e^{i(\omega\tau+\phi_i)} \quad 0 \le \xi_i \le l_i \tag{2.46.1}$$

$$H_i(\xi_i, x_0; \omega) = \frac{\sqrt{-1}}{2\gamma}[f_i(\xi_i; \omega) + f_i^{-1}(\xi_i; \omega) R_{ir}] \prod_{k=i}^{1} T_k \quad \phi_i = \angle H_i \tag{2.46.2}$$

One of the main advantages of this approach is its systematic formulation resulting in a recursive computational algorithm which can be implemented into highly efficient computer codes consuming less computer resources. This systematic approach also allows modular formulation which can be easily expandable to include additional subspans with very minor alteration to the existing formulation. Another significant advantage of the present wave-based approach to the forced response analysis of a multi-span string is that the eigensolutions of the system is not required as a priori as in the method of normal mode expansion which assumes the forced response solution in terms of an infinite series of the system eigenfunctions - truncated later for numerical computations. However, exact eigensolutions are often difficult to obtain particularly for non-self-adjoint systems, and also approximated eigensolutions can result in large error. In contrast, the current analysis technique renders closed-form transfer functions from which exact frequency response solutions can be obtained.

3. Fourth order systems

The analysis techniques described by using the vibration of a string can be applied to the transverse vibration of a beam of which equation of motion is typically a fourth order partial differential equation. Denoting X and t as the spatial and temporal variables, respectively, the equation of motion governing the transverse displacement $W(X,t)$ of a damped uniform Euler-Bernoulli beam of length L subjected to an external load $P(X,t)$ is

$$m\frac{\partial^2 W}{\partial t^2} + C_e\frac{\partial W}{\partial t} + EI\frac{\partial^4 W}{\partial X^4} = P(X,t) \tag{3.1}$$

where m denotes the mass per unit length, EI the flexural rigidity, C_e the external damping coefficient of the beam. With introduction of the following non-dimensional variables and parameters

$$w = W/L,\ x = X/L,\ \tau = t/t_0,\ t_0 = mL^4/EI\ \ c_e = C_e t_0/m,\ p(x,t) = P(X,t)L^3/EI \tag{3.2}$$

the equation of motion takes the non-dimensional form of

$$\ddot{w} + c_e\dot{w} + w^{(4)} = p(x,t) \quad (0 \le x \le 1) \tag{3.3}$$

Applying the Laplace transform to Eq. (3.3) yields

$$s^2\overline{w}(x;s) + c_e s\overline{w}(x;s) + \overline{w}^{(4)}(s;x) = \overline{p}(x;s) \tag{3.4}$$

Letting $\overline{p}(x;s) = 0$, the homogeneous wave solution of Eq. (3.4) can be assumed as:

$$\overline{w}(x;s) = Ce^{i\gamma x} \tag{3.5}$$

where γ is the non-dimensional wavenumber normalized against span length L. Applying the above wave solution to Eq. (3.4) gives the frequency equation of the problem

$$\gamma^4 = -(s^2 + c_e s) \tag{3.6}$$

from which the general wave solution can be found as the sum of four constituent waves

$$\overline{w}(x;s) = C_a^+ e^{-i\gamma x} + C_a^- e^{i\gamma x} + C_b^+ e^{-\gamma x} + C_b^- e^{\gamma x} \tag{3.7}$$

where the coefficient C represents the amplitude of each wave with its traveling direction indicated by the superscript; plus (+) and minus (-) signs denote the wave's traveling directions with respect to the x-coordinate. The subscripts a and b signify the spatial wave motion of the same type traveling in the opposite direction. Note that γ is complex valued in general. The general wave solution in Eq. (3.7) may be re-expressed in vector form by grouping the wave components (wave packet) traveling in the same direction; i.e.,

$$\mathbf{C}^+ = \begin{Bmatrix} C_a^+ \\ C_b^+ \end{Bmatrix} \quad \mathbf{C}^- = \begin{Bmatrix} C_a^- \\ C_b^- \end{Bmatrix} \tag{3.8}$$

and then

$$\overline{w}(x;s) = [1 \quad 1][\mathbf{f}(x;s)\mathbf{C}^+ + \mathbf{f}^{-1}(x;s)\mathbf{C}^-] \tag{3.9}$$

where $\mathbf{f}(x;s)$ is the diagonal *field transfer matrix* (Mace, 1984) given by

$$\mathbf{f}(x;s) = \begin{bmatrix} e^{-i\gamma x} & 0 \\ 0 & e^{-\gamma x} \end{bmatrix} \tag{3.10}$$

which relates the wave amplitudes by

$$\mathbf{C}^+(x + x_0) = \mathbf{f}\mathbf{C}^+(x_0) \quad \text{or} \quad \mathbf{C}^-(x + x_0) = \mathbf{f}^{-1}\mathbf{C}^-(x_0) \tag{3.11}$$

3.1 Wave reflection and transmission matrices

For a wave packet with multiple wave components, the rates of wave reflection and transmission at a point discontinuity can be found in terms of the *wave reflection matrix* **r** and *wave transmission matrix* **t**, in the same manner described in Section 2.2. When the flexural wave packet in Eq. (3.8) travels along a beam and is incident upon a kinetic constraint (ξ=0) which consists of, for example, transverse (K_t) and rotational (K_r) springs and transverse damper (C_t), **r** and **t** at the discontinuity can be found by applying the geometric continuity kinetic equilibrium conditions at ξ=0; i.e., with reference to Fig. 7, one can establish the following matrix equations

$$\begin{bmatrix} 1 & 1 \\ -i & -1 \end{bmatrix}\mathbf{C}^+ + \begin{bmatrix} 1 & 1 \\ i & 1 \end{bmatrix}\mathbf{rC}^+ = \begin{bmatrix} 1 & 1 \\ -i & -1 \end{bmatrix}\mathbf{tC}^+ \tag{3.12.1}$$

$$\begin{bmatrix} 1 & -1 \\ i & -1 \end{bmatrix}\mathbf{C}^+ + \begin{bmatrix} 1 & -1 \\ -i & 1 \end{bmatrix}\mathbf{rC}^+ = \begin{bmatrix} 1 - ik_r/\gamma & -1 - k_r/\gamma \\ i + (k_t + c_t s)/\gamma^3 & -1 + (k_t + c_t s)/\gamma^3 \end{bmatrix}\mathbf{tC}^+ \tag{3.12.2}$$

where $k_t = K_t L^3/EI$, $k_r = K_r L/EI$, and $c_t = C_t t_0/m$. Solving the above equations gives the local wave reflection and transmission matrices as:

$$\mathbf{r} = \frac{1}{2}\begin{bmatrix} (1-i)(\alpha - \beta i) & -(1+i)(\alpha+\beta) \\ -(1-i)(\alpha+\beta) & (1+i)(\alpha+\beta i) \end{bmatrix} \tag{3.13.1}$$

$$\mathbf{t} = \frac{1}{2}\begin{bmatrix} 2-(1-i)\alpha-(1+i)\beta & (1+i)(\alpha-\beta) \\ (1-i)(\alpha-\beta) & 2-(1+i)\alpha-(1-i)\beta \end{bmatrix} \tag{3.13.2}$$

$$\alpha = \frac{k_r + c_r s}{k_r + c_r s + (2+2i)\gamma} \qquad \beta = \frac{k_t + c_t s}{k_t + c_t s - (2-2i)\gamma^3}$$

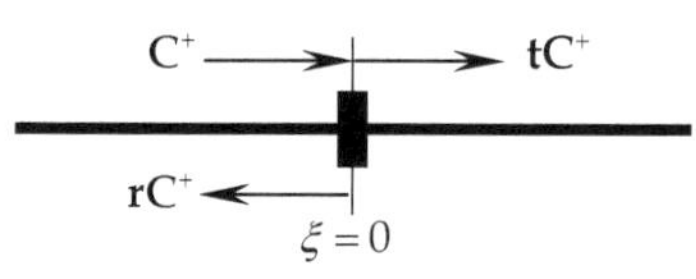

Fig. 7. Wave reflection and transmission at a discontinuity.

However, as previously discussed in Section 2.2, when the wave packet is incident upon a series of discontinuities along its traveling path, it is more computationally efficient to employ the concepts of *global wave reflection* and *transmission matrices*. These matrices relate the amplitudes of incoming and outgoing waves at a point discontinuity. When compared to the string problem, the only difference in formulating these matrices for the beam problem is to use vectors and matrices instead of single coefficients. Therefore, with reference to Fig. 2, let $\mathbf{R}_{ir}$ as the global wave reflection matrix which relates the amplitudes of negative- and positive-traveling waves on the right side of discontinuity i such that

$$\mathbf{C}_{ir}^- = \mathbf{R}_{ir}\mathbf{C}_{ir}^+ \tag{3.14}$$

Since $\mathbf{C}_{ir}^{-} = \mathbf{f}_i \mathbf{R}_{(i+1)l} \mathbf{C}_{ir}^{+}$, one can find $\mathbf{R}_{ir}$ in terms of the global wave reflection matrix on the left side of discontinuity i+1; i.e.,

$$\mathbf{R}_{ir} = \mathbf{f}_i \mathbf{R}_{(i+1)l} \mathbf{f}_i \tag{3.15}$$

In addition, by combining the following wave equations at discontinuity i

$$\mathbf{C}_{ir}^{+} = \mathbf{t}_i \mathbf{C}_{il}^{+} + \mathbf{r}_i \mathbf{C}_{ir}^{-} \quad \mathbf{C}_{il}^{-} = \mathbf{t}_i \mathbf{C}_{ir}^{-} + \mathbf{r}_i \mathbf{C}_{il}^{+} \tag{3.16-17}$$

the relationship between the global wave reflection matrices on the left and right sides of discontinuity i can be found as

$$\mathbf{R}_{il} = \mathbf{r}_i + \mathbf{t}_i (\mathbf{R}_{ir}^{-1} - \mathbf{r}_i)^{-1} \mathbf{t}_i \tag{3.18}$$

$\mathbf{R}_{ir}$ and $\mathbf{R}_{il}$ progressively expand to include all the global wave reflection matrices of intermediate discontinuities along the beam before terminating its expansion at the boundaries. Since incident waves are only reflected at the boundaries, one can find the following wave equations

$$\mathbf{C}_{1}^{+} = \mathbf{r}_1 \mathbf{C}_{1}^{-} \quad \mathbf{C}_{n}^{-} = \mathbf{r}_n \mathbf{C}_{n}^{+} \tag{3.19.1-2}$$

where $\mathbf{r}$ can be found by imposing the force and moment equilibrium conditions at the boundary; e.g., for the same kinetic constraint previously considered

$$\mathbf{r} = \begin{bmatrix} \gamma - ik_r & -\gamma - k_r \\ -i\gamma^3 - (k_t + c_t s) & \gamma^3 - (k_t + c_t s) \end{bmatrix}^{-1} \begin{bmatrix} \gamma + ik_r & -\gamma + k_r \\ i\gamma^3 - (k_t + c_t s) & -\gamma^3 - (k_t + c_t s) \end{bmatrix} \tag{3.20}$$

Now, to determine the global wave transmission matrix $\mathbf{T}_i$, define

$$\mathbf{C}_{ir}^{+} = \mathbf{T}_i \mathbf{C}_{(i-1)r}^{+} \tag{3.21}$$

Rewriting Eq. (3.16) by applying $\mathbf{C}_{il}^{+} = \mathbf{f}_{(i-1)} \mathbf{C}_{(i-1)r}^{+}$ and $\mathbf{C}_{ir}^{-} = \mathbf{R}_{ir} \mathbf{C}_{ir}^{+}$, and then comparing it with Eq. (3.21), one can find that

$$\mathbf{T}_i = (\mathbf{I}_{2\times 2} - \mathbf{r}_i \mathbf{R}_{ir})^{-1} \mathbf{t}_i \mathbf{f}_{(i-1)} \tag{3.22}$$

where $\mathbf{I}_{2\times 2}$ is the 2×2 identity matrix.

3.2 Free response analysis

The global reflection and transmission matrices of waves traveling along a multi-span beam are now combined with the field transfer matrix to analyze the free vibration of the beam. With reference to Fig. 2, where $C_i^{\pm}$, R_i, and f_i are now replaced by $\mathbf{C}_i^{\pm}$, $\mathbf{R}_i$, and $\mathbf{f}_i$, respectively, at the left boundary

$$\mathbf{C}_{1}^{-} = \mathbf{R}_{1r} \mathbf{C}_{1}^{+} \tag{3.23}$$

However, due to Eq. (3.19), it can be found that

$$(\mathbf{r}_1\mathbf{R}_{1r} - \mathbf{I}_{2\times2})\mathbf{C}_1^+ = 0 \tag{3.24}$$

Applying the condition for non-trivial solutions to the above matrix equation, one can obtain the following characteristic equation in terms of the Laplace variable s

$$F(s) = \mathrm{Det}[\mathbf{r}_1\mathbf{R}_{1r} - \mathbf{I}_{2\times2}] = 0 \tag{3.25}$$

By applying a standard root search technique (e.g., Newton-Raphson method or secant method) to Eq. (3.25), one can obtain the natural frequencies of the multi-span beam.
The mode shapes of the multi-span beam can be systematically found by relating wave amplitudes between two adjacent subspans, in the same way described in Section 2.3. Defining ξ_i as the local coordinate in span i, the transverse displacement of any point in span i can be found as

$$\bar{w}_i(\xi_i) = [1 \quad 1][\mathbf{f}_i(\xi_i)\mathbf{C}_i^+ + \mathbf{f}_i^{-1}(\xi_i)\mathbf{C}_i^-] \tag{3.26}$$

However, due to Eq. (3.14)

$$\bar{w}_i(\xi_i) = [1 \quad 1][\mathbf{f}_i(\xi_i) + \mathbf{f}_i^{-1}(\xi_i)\mathbf{R}_{ir}]\mathbf{C}_i^+ \tag{3.27}$$

Since $\mathbf{C}_i^+ = \mathbf{T}_i\mathbf{C}_{i-1}^+$ from Eq. (3.21),

$$\bar{w}_i(\xi_i) = [1 \quad 1][\mathbf{f}_i(\xi_i) + \mathbf{f}_i^{-1}(\xi_i)\mathbf{R}_{ir}]\mathbf{T}_i\mathbf{C}_{i-1}^+ \tag{3.28}$$

Assume a wave packet originates and starts traveling from the leftmost boundary of the beam. By successively applying the global transmission matrix of each discontinuity on the way up to the first span, the mode shape of span i can be found in terms of wave amplitude $\mathbf{C}_1^+$; i.e.,

$$\bar{w}_i(\xi_i) = [1 \quad 1][\mathbf{f}_i(\xi_i) + \mathbf{f}_i^{-1}(\xi_i)\mathbf{R}_{ir}]\prod_{j=i}^{1}\mathbf{T}_j\mathbf{C}_1^+ \quad \mathbf{T}_1 = \mathbf{I}_{2\times2} \quad 0 \le \xi_i \le l_i \tag{3.29}$$

	Discontinuity					
Constraint	1	2	3	4	5	6
Location	0	0.15	0.35	0.6	0.82	1
k_t	∞	3,000	2,500	1,500	3,500	∞
k_r	0	250	150	100	300	∞
c_t	0	0	0	0	0	0

Table 1. Nondimensional system parameter used for Fig. 8.

where note that the amplitude ratio between the two wave components C_a^+ and C_b^+ can be found from Eq. (3.24). For example, shown in Fig. 8 are the first three mode shapes of a uniformly damped five-span beam with system parameters specified in Table 1. Once the wave reflection and transmission matrices at each discontinuity and the boundary are determined, one can apply Eq. (3.25) to find the first three natural wavenumbers γ_1=10.294, γ_2=12.038, and γ_3=14.148, from which the damped natural frequencies of the beam can be determined by using Eq. (3.6). It should be noted from the computational point of view that the present wave approach always results in operationg matrices of a fixed size regardless of the number of subspans. However if the classical method of separation of variables is applied to solve a multi-span beam problem, the size of matrix that determines the eigensolutions of the problem increases as the number of subspans increases, which may cause strenuous computations associated with large-order matrices.

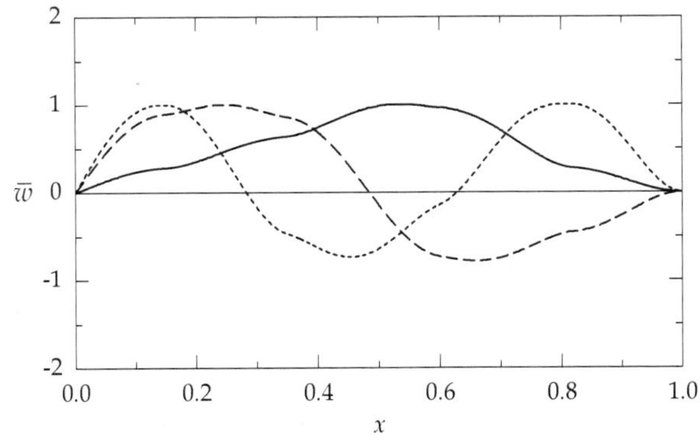

Fig. 8. First three mode shapes obtained from Eq. (3.25). The solid, dashed, and dotted curves represent the 1st, 2nd, and 3rd modes, respectively.

3.3 Transfer function analysis

Consider a multi-span beam subjected to an external force applied at $x=x_0$ where x_0 is located in subspan m. Let $\mathbf{C}^{\pm}$ be the amplitudes of the waves rise in sub-span n as a result of injected waves due to the applied force, and also assume that $\mathbf{C}^{\pm}$ satisfy all the continuity conditions at intermediate discontinuities and boundary conditions of the multi-span beam system. The transverse displacement $\bar{w}_n(x;s)$ of span n can be expressed in wave form

$$\bar{w}_n(x;s) = [1 \quad 1][\mathbf{f}(x;s)\mathbf{C}^{+} + \mathbf{f}^{-1}(x;s)\mathbf{C}^{-}] \tag{3.30}$$

Now, in order to determine the actual wave amplitudes $\mathbf{C}^{\pm}$, consider the multi-span beam with arbitrary supports and boundary conditions under a concentrated applied force of $\bar{p}(s)\delta(x-x_0)$, where $\bar{p}(s)$ is the Laplace transform of $p(\tau)$, as schematically depicted in Fig. 5 with $C_i^{\pm}$, $D_i^{\pm}$, and R_i replaced by $\mathbf{C}_i^{\pm}$, $\mathbf{D}_i^{\pm}$, and $\mathbf{R}_i$, respectively. Since the waves injected at $x=x_0$ travel in both directions, a new set of local coordinates $\{\xi, \xi^*\}$ is defined such that the waves traveling towards each end of the beam are measured positive as indicated in the Fig. 5. Let $\mathbf{C}_i^{\pm}$ and $\mathbf{D}_i^{\pm}$ be the amplitudes of the waves traveling within subspan i in the region $x<x_0$ and $x\geq x_0$, respectively. The discontinuity in the shear force at $x=x_0$ can be expressed in term of the difference in amplitudes between the incoming and outgoing waves at the discontinuity such that

$$\mathbf{D}_1^{+} - \mathbf{C}_1^{-} = \mathbf{v} \quad \mathbf{C}_1^{+} - \mathbf{D}_1^{-} = \mathbf{v} \tag{3.31}$$

where $\mathbf{v}$ is the wave vector injected by the applied force which can be determined by the geometric and kinetic continuity conditions at $\xi=0$ as

$$\mathbf{v} = \frac{\overline{p}(s)}{4\gamma^3}\begin{Bmatrix}\sqrt{-1}\\ 1\end{Bmatrix} \tag{3.32}$$

Applying the global wave reflection matrices on each side of discontinuity 1 gives

$$\mathbf{C}_1^- = \mathbf{R}_{1r}\mathbf{C}_1^+ \quad \mathbf{D}_1^- = \mathbf{R}_{1r}^*\mathbf{D}_1^+ \tag{3.33.1-2}$$

Combining Eqs. (3.31) and (3.33), one can find the wave amplitudes that give rise at the point of loading

$$\mathbf{C}_1^+ = \mathbf{T}_1\mathbf{v} \quad \mathbf{T}_1 = (\mathbf{I} - \mathbf{R}_{1r}^*\mathbf{R}_{1r})^{-1}(\mathbf{I} + \mathbf{R}_{1r}^*) \tag{3.34}$$

$$\mathbf{D}_1^+ = \mathbf{T}_1^*\mathbf{v} \quad \mathbf{T}_1^* = (\mathbf{I} - \mathbf{R}_{1r}\mathbf{R}_{1r}^*)^{-1}(\mathbf{I} + \mathbf{R}_{1r}) \tag{3.35}$$

$\mathbf{T}_1$ and $\mathbf{T}_1^*$ in the above equations can be considered as the generalized wave transmission matrices that characterize the transmissibility of the waves injected by the applied force in the region $x<x_0$ and $x\geq x_0$, respectively. It is evident that $\mathbf{T}_1$ and $\mathbf{T}_1^*$ are different unless the system is symmetric about $x=x_0$. Applying the results in Eqs. (3.34) and (3.35) to Eq. (3.30), the transverse displacement of any point in span 1 on either side of the point of loading can be found. For example, for span 1 in the region $x<x_0$

$$\overline{w}_1(\xi_1,x_0;s) = [1 \quad 1][\mathbf{f}_1(\xi_1;s) + \mathbf{f}_1^{-1}(\xi_1;s)\mathbf{R}_{1r}]\mathbf{T}_1\mathbf{v} \quad 0 \leq \xi_1 \leq l_1 \tag{3.36}$$

In the same manner as for free response analysis, the forced response solution in Eq. (3.36) can be generalized for a multi-span beam by applying the global wave transmission matrix. For $x<x_0$, $\mathbf{C}_1^+ = \mathbf{T}_1\mathbf{v}$ and $\mathbf{C}_{i+1}^+ = \mathbf{T}_i\mathbf{C}_i^+$. Therefore one can progressively expand the solution in Eq. (3.36) until the expansion terminates at the leftmost boundary. Thus, the transverse displacement of any point within span i in the region $x<x_0$ due to external loading at $x=x_0$ can be determined from

$$\overline{w}_i(\xi_i,x_0;s) = \frac{\overline{p}(s)}{4\gamma^3}[1 \quad 1][\mathbf{f}_i(\xi_i;s) + \mathbf{f}_i^{-1}(\xi_i;s)\mathbf{R}_{ir}]\prod_{k=i}^{1}\mathbf{T}_k\begin{Bmatrix}\sqrt{-1}\\ 1\end{Bmatrix} \quad 0 \leq \xi_i \leq l_i \tag{3.37}$$

One can find the same expression in the region $x\geq x_0$, but in terms of $\mathbf{R}_{ir}^*$ and $\mathbf{T}_k^*$ instead. Note that the ratio $G_i(\xi,x_0;s) = \overline{w}(\xi_i,x_0;s)/\overline{p}(s)$ in Eq. (3.37) is the transfer function governing the forced response of anypoint in span i due to the point loading at $x=x_0$. The Laplace inversion of $G_i(\xi,x_0;s)$ is the Green's function of the problem. Therefore, denoting L^{-1} as the inverse Laplace transform operator, the forced response at any point within any span of the multi-span beam can be found from; e.g., for span i in the region $x<x_0$,

$$w_i(\xi_i,x_0;\tau) = \int_0^t G_i(\xi_i,x_0;\tau-\overline{\tau})p(\tau)d\overline{\tau} \quad (i=1,2,\cdots,m) \tag{3.38}$$

$$G_i(\xi_i,x_0;\tau) = L^{-1}\left[\frac{1}{4\gamma^3}[1 \quad 1][\mathbf{f}_i(\xi_i;s) + \mathbf{f}_i^{-1}(\xi_i;s)\mathbf{R}_{ir}]\prod_{k=i}^{1}\mathbf{T}_k\begin{Bmatrix}\sqrt{-1}\\ 1\end{Bmatrix}\right] \quad 0 \leq \xi_i \leq l_i \tag{3.39}$$

The exact Laplace inversion of $G_i(\xi, x_0; s)$ in close form is prohibitively difficult for the beam problems considered in this study, in particular multi-span beams with non-classical intermediate supports and boundary conditions. One may have to resort to the numerical inversion of Laplace transforms. Shown in Fig. 9 is the unit impuse response curve of a non-uniformly damped three-span beam, whose system parameters are listed in Table 2, sampled at x=0.15 and x=0.825 for x_0= 0.4. The unit impulse response curve obtained from the method of normal mode expansion is also shown for comparison. For the results obtained from the method of normal mode expansion, the mode shape function of $\sin n\pi x$ (n=1, ..., 10) are used for the expansion and the constraint forces at the intermediate supports are represented by using the Dirac-Delta function. The resulting set of modal equations are numerically integrated using a fixed-step Runge-Kutta algorithm since the beam is nonproportionally damped hence the modal equations cannot be decoupled. It can be seen that both results show an excellent agreement.

	Discontinuity				
Constraint	3	2	1=1*	2	3
Location	0	0.15	0.4	0.65	1
k_t	∞	2,000	0	3,000	∞
k_r	0	0	0	0	0
c_t	0	15	0	10	0

Table 2. Nondimensional system parameter used for Fig. 9, where c_e=0.

Regarding the numerical Laplace inversion to obtain the transient response of a Euler-Bernoulli beam model due to an impulse, it should be noted that there exists a singular behavior of the response for small values of time; i.e., the initial response. This is not due to the algorithm of the numerical Laplace inversion, but due to the inherent deficiency of the classical Euler-Bernoulli beam theory that neglects the effects of rotary inertia and shear

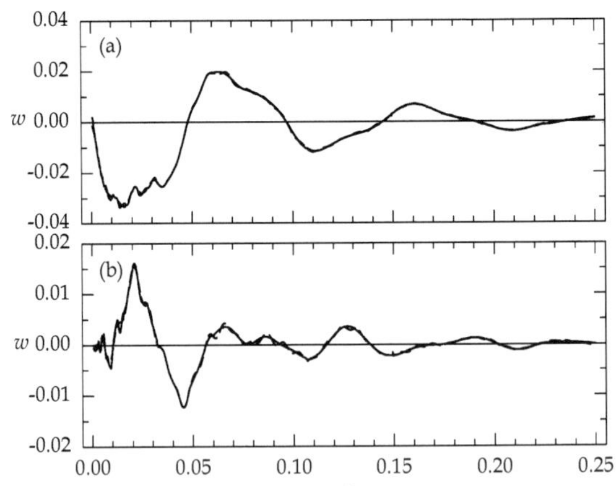

Fig. 9. Unit impulse response of the three-span beam sampled at (a) x=0.15 and (b) x=0.825. The solid and dashed curves represent the results from the present analysis and the method of normal mode expansion, respectively.

deformations. According to this simple beam theory, a wave of very high frequency (or very short wavelength) would be predicted instantaneously at remote locations, which is physically unacceptable since this can be interpreted as that such a wave travels at an infinity velocity. This abnormal behavior of the Euler-Bernoulli beam can be readily verified by observing the phase velocity c of the dispersion equation (Eq. (3.6) with $s=i\omega$ and $\omega=c\gamma$) continues to increase with increasing γ. Therefore, for a meaningful initial transient response to an impulsive load, one must employ Rayleigh or Timoshenko beam models. Studies on the wave reflection and transmission behavior at various point discontinuities and boundaries using a Timoshenko beam model can be found in Refs. (Tan & Kang, 1999; Mei and Mace, 2005).

If $p(\tau)=p_0 e^{i\omega\tau}$; i.e., a harmonic forcing function, the steady-state response of the problem can be readily found in terms of the complex frequency function defined as

$$H(\xi_i, x_0; \omega) = G(\xi_i, x_0; s)\,|_{s=i\omega} \tag{3.40}$$

	Discontinuity							
Constraint	4	3	2	1=1*	2*	3*	4*	5*
Location	0	0.12	0.26	0.39	0.56	0.69	0.84	1
k_t	5,000	4,000	3,500	0	3,000	5,000	∞	0
k_r	100	0	100	0	100	50	0	0
c_t	0.5	0.1	0	0	0	0.1	0.1	0.2

Table 3. Nondimensional system parameter used for Fig. 8, where c_e=0.

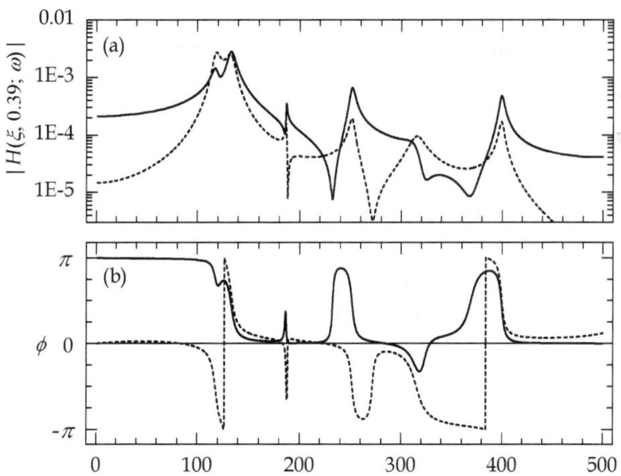

Fig. 10. Frequency response of the six-span beam due to harmonic excitation applied at x=0.39: (a) amplitudes sampled at x=0.325 (solided curve) and x=0.92 (dashed curve); and (b) corresponding phase angle.

Therefore the frequency response at any point within any subspan of the beam can be obtained by; e.g., for span i in the region $x<x_0$

$$w_i(\xi_i,x_0;\tau)=|H_i(\xi_i,x_0;\omega)|\,p_0e^{i(\omega\tau+\phi_i)}\quad 0\le\xi_i\le l_i \tag{3.41}$$

$$H_i(\xi_i,x_0;\omega)=\frac{1}{4\gamma^3}[1\quad 1][\mathbf{f}_i(\xi_i;\omega)+\mathbf{f}_i^{-1}(\xi_i;\omega)\mathbf{R}_{i_r}]\prod_{k=i}^{1}\mathbf{T}_k\begin{Bmatrix}\sqrt{-1}\\1\end{Bmatrix}\quad \phi_i=\angle H_i \tag{3.42}$$

Shown in Fig.10 is the frequency response sampled at x=0.325 and x=0.92 for the non-uniformly damped six-span beam, whose system parameters are listed in Table 3, under harmonic excitation at x=0.39.

4. Sixth order systems

The analysis techniques described in the previous sections can be applied to the vibrations of higher order, one-dimensional, distributed parameter systems. For example, the in-plane vibration of a planar curved beam is governed by a sixth order partial differential equation. Consider a thin planar curved bam. Neglecting the effects of rotary inertia and shear deformations, the coupled equations of motion governing the flexural displacement W and the tangential displacement U of the centroidal axis are

$$\frac{EI}{R^3}\frac{\partial^3}{\partial\theta^3}\left(U-\frac{\partial W}{\partial\theta}\right)-\frac{EA}{R}\left(W+\frac{\partial U}{\partial\theta}\right)=\rho AR\frac{\partial^2W}{\partial t^2} \tag{4.1.1}$$

$$\frac{EI}{R^3}\frac{\partial^2}{\partial\theta^2}\left(U-\frac{\partial W}{\partial\theta}\right)+\frac{EA}{R}\frac{\partial}{\partial\theta}\left(W+\frac{\partial U}{\partial\theta}\right)=\rho AR\frac{\partial^2U}{\partial t^2} \tag{4.1.2}$$

where E denotes the Young's modulus, I the second moment of inertia of the cross-section about the centroid, θ the angular coordinate, R the constant radius of curvature for the given range of angle θ, A the cross-sectional area, ρ the mass density, and t the time variable. Details of deriving these equations of motion can be found in Ref. (Graff, 1975). With the following non-dimensional variables and parameters:

$$u=\frac{U}{R}\quad w=\frac{W}{R}\quad \tau=\frac{t}{t_0}\quad t_0=R^2\sqrt{\frac{\rho A}{EI}}\quad k^2=\frac{I}{AR^2} \tag{4.2}$$

where t_0 is a characteristic time constant and k is the curvature parameter, Eq. (4.1) takes the following non-dimensional form in the Laplace domain

$$k^2\frac{\partial^3}{\partial\theta^3}\left(\overline{u}(\theta;s)-\frac{\partial\overline{w}(\theta;s)}{\partial\theta}\right)-\left(\overline{w}(\theta;s)+\frac{\partial u(\theta;s)}{\partial\theta}\right)=k^2s^2\overline{w}(\theta;s) \tag{4.3.1}$$

$$k^2\frac{\partial^2}{\partial\theta^2}\left(\overline{u}(\theta;s)-\frac{\partial\overline{w}(\theta;s)}{\partial\theta}\right)+\frac{\partial}{\partial\theta}\left(\overline{w}(\theta;s)+\frac{\partial\overline{u}(\theta;s)}{\partial\theta}\right)=k^2s^2\overline{u}(\theta;s) \tag{4.3.2}$$

where s represents the Laplace variable. In order to determine the general wave solutions for the equations of motion in Eq. (4.3), the following form of harmonic wave solutions are first assumed

$$\begin{Bmatrix}\overline{w}(\theta;s)\\\overline{u}(\theta;s)\end{Bmatrix}=\begin{Bmatrix}C_w\\C_u\end{Bmatrix}e^{i\gamma\theta} \tag{4.4}$$

where γ denotes the wavenumber, normalized against R, of the wave traveling along the centroidal axis. Substituting the above wave solutions into Eq. (4.3) leads to

$$\begin{bmatrix} k^2(\gamma^4+s^2)+1 & i\gamma(1+\gamma^2k^2) \\ i\gamma(1+\gamma^2k^2) & -k^2(\gamma^2+s^2)-\gamma^2 \end{bmatrix} \begin{Bmatrix} C_w \\ C_u \end{Bmatrix} = 0 \tag{4.5}$$

Since the determinant of the matrix in Eq. (4.5) must vanish for nontrivial solutions, one can obtain following frequency equation

$$\gamma^6 + (k^2s^2-2)\gamma^4 + [1+(1+k^2)s^2]\gamma^2 + (k^2s^2+1)s^2 = 0 \tag{4.6}$$

The six roots (essentially three complex conjugates, namely $\pm\gamma_n(s)$, n=1 ,2, 3) of Eq. (4.6) can be obtained in a closed form by transforming Eq. (4.6) into a cubic equation. Each conjugate pair is the wavenumber of the constituent wave components corresponding to each of the two wave modes of the curved beam. Therefore, it can be found that

$$\bar{w}(\theta;s) = \sum_{n=1}^{3} (C_n^+ e^{-i\gamma_n\theta} + C_n^- e^{i\gamma_n\theta}) \tag{4.7.1}$$

$$\bar{u}(\theta;s) = \sum_{n=1}^{3} \alpha_n (C_n^+ e^{-i\gamma_n\theta} - C_n^- e^{i\gamma_n\theta}) \tag{4.7.1}$$

where the coefficient C represents the amplitude of each flexural wave component with its traveling direction indicated by the +/− signs. The subscript n signifies the spatial wave motion of the same type traveling in the opposite direction. α_n is the amplitude ratio between the flexural and tangential waves which can be found from Eq. (4.5) that

$$\alpha_n = \frac{i\gamma_n(k^2\gamma_n^2+1)}{(1+k^2)\gamma_n^2 + k^2s^2} \tag{4.8}$$

The above wave solutions may be recast in vector form by grouping the waves traveling in the same direction in a wave packet; i.e., $\mathbf{C}^+ = \{C_1^+ \quad C_2^+ \quad C_3^+\}^T$ and $\mathbf{C}^- = \{C_1^- \quad C_2^- \quad C_3^-\}^T$, then

$$\bar{w}(\theta;s) = \mathbf{u}[\mathbf{f}\mathbf{C}^+ + \mathbf{f}^{-1}\mathbf{C}^-] \tag{4.9.1}$$

$$\bar{u}(\theta;s) = \boldsymbol{\alpha}[\mathbf{f}\mathbf{C}^+ - \mathbf{f}^{-1}\mathbf{C}^-] \tag{4.9.2}$$

where $u = [1 \quad 1 \quad 1]$, $\boldsymbol{\alpha} = [\alpha_1 \quad \alpha_2 \quad \alpha_3]$, and $\mathbf{f}$ is the field transfer matrix defined by

$$\mathbf{f}(\theta;s) = \begin{bmatrix} e^{-i\gamma_1\theta} & 0 & 0 \\ 0 & e^{-i\gamma_2\theta} & 0 \\ 0 & 0 & e^{-i\gamma_3\theta} \end{bmatrix} \tag{4.10}$$

One now can see that the analysis techniques described in the previous sections for the multi-span beam can be applied to a multi-span curved beam in the same manner to determine the local wave reflection and transmission matrices, global wave reflection and transmission matrices, and the transfer function (Kang et al., 2003). The only difference is the size of matrix, three by three for the curved beam. It should be also noted that the present analysis techniques are applicable to one-dimensional distributed parameter systems involving other types of discontinuities such as geometric/material changes and cracks, as long as the properties of those discontinuities can be characterized by wave reflection and transmission coefficients or matrices. In what follows, an example of free vibration analysis of a three-span curved beam with curvature changes is presented.

4.1 Wave reflection and transmission at a curvature change

Consider a curved beam with two subspans of different curvatures joined at ξ=0. Assuming that the wavenumbers of the waves traveling in each subspan are γ_n and γ'_n (n=1, 2, 3), the geometric continuity conditions at ξ=0 give

$$\begin{bmatrix} 1 & 1 & 1 \\ \alpha_1 & \alpha_2 & \alpha_3 \\ \alpha_1 - i\gamma_1 & \alpha_2 - i\gamma_2 & \alpha_3 - i\gamma_3 \end{bmatrix}\mathbf{C}^+ + \begin{bmatrix} 1 & 1 & 1 \\ -\alpha_1 & -\alpha_2 & -\alpha_3 \\ -\alpha_1 + i\gamma_1 & -\alpha_2 + i\gamma_2 & -\alpha_3 + i\gamma_3 \end{bmatrix}\mathbf{rC}^+ = \begin{bmatrix} 1 & 1 & 1 \\ \alpha'_1 & \alpha'_2 & \alpha'_3 \\ \alpha'_1 - i\gamma'_1 & \alpha_2 - i\gamma'_2 & \alpha'_3 - i\gamma'_3 \end{bmatrix}\mathbf{tC}^+ \tag{4.11.1}$$

and the kinetic continuity conditions give

$$\begin{bmatrix} 1 - i\gamma_1\alpha_1 & 1 - i\gamma_2\alpha_2 & 1 - i\gamma_3\alpha_3 \\ -\gamma_1^2(i\gamma_1 + \alpha_1) & -\gamma_2^2(i\gamma_2 + \alpha_2) & -\gamma_3^2(i\gamma_3 + \alpha_3) \\ \gamma_1(\gamma_1 - i\alpha_1) & \gamma_2(\gamma_2 - i\alpha_2) & \gamma_3(\gamma_3 - i\alpha_3) \end{bmatrix}\mathbf{C}^+ + \begin{bmatrix} 1 - i\gamma_1\alpha_1 & 1 - i\gamma_2\alpha_2 & 1 - i\gamma_3\alpha_3 \\ \gamma_1^2(i\gamma_1 + \alpha_1) & \gamma_2^2(i\gamma_2 + \alpha_2) & \gamma_3^2(i\gamma_3 + \alpha_3) \\ \gamma_1(\gamma_1 - i\alpha_1) & \gamma_2(\gamma_2 - i\alpha_2) & \gamma_3(\gamma_3 - i\alpha_3) \end{bmatrix}\mathbf{rC}^+ = \begin{bmatrix} 1 - i\gamma'_1\alpha'_1 & 1 - i\gamma'_2\alpha'_2 & 1 - i\gamma'_3\alpha'_3 \\ -\gamma_1'^2(i\gamma'_1 + \alpha'_1) & -\gamma_2'^2(i\gamma'_2 + \alpha'_2) & -\gamma_3'^2(i\gamma'_3 + \alpha'_3) \\ \gamma'_1(\gamma'_1 - i\alpha'_1) & \gamma'_2(\gamma'_2 - i\alpha'_2) & \gamma'_3(\gamma'_3 - i\alpha'_3) \end{bmatrix}\mathbf{tC}^+ \tag{4.11.2}$$

$$\alpha'_n = \frac{i\gamma'_n(k'^2\gamma_n'^2 + 1)}{(1 + k'^2)\gamma_n'^2 + k'^2 s^2} \tag{4.11.3}$$

The local wave reflection and transmission matrices can be found by solving the above equations. The curved beam system shown in Fig. 11 has three subspans of equal span angle of 60° but different radius of curvature. Both ends are clamped. The natural frequencies of the system can be obtained from

$$F(s) = \mathrm{Det}[\mathbf{r}_1\mathbf{R}_{1r} - \mathbf{I}_{3\times3}] = 0 \tag{4.12}$$

where $\mathbf{r}_1$ is the wave reflection matrix at the clamed boundary, which is

$$\mathbf{r}_1 = \begin{bmatrix} -1 & -1 & -1 \\ -\alpha_1 & -\alpha_2 & -\alpha_3 \\ \alpha_1 - i\gamma_1 & \alpha_2 - i\gamma_2 & \alpha_3 - i\gamma_3 \end{bmatrix}^{-1} \begin{bmatrix} -1 & -1 & -1 \\ -\alpha_1 & -\alpha_2 & -\alpha_3 \\ \alpha_1 - i\gamma_1 & \alpha_2 - i\gamma_2 & \alpha_3 - i\gamma_3 \end{bmatrix} \quad (4.13)$$

and $\mathbf{R}_{1r}$ is the global wave reflection matrix that includes the effects of the curvature changes and the other clamped boundary. The natural frequencies are compared in Table 4 with the results obtained by the finite element method (FEM) (Maurizi et al., 1993; Krishnan and Suresh, 1998) and the cell discretization method (CDM) (Maurizi et al., 1993). Note that the natural frequencies obtained from the present wave analysis are exact since both propagating and attenuating wave components are considered in the formulation.

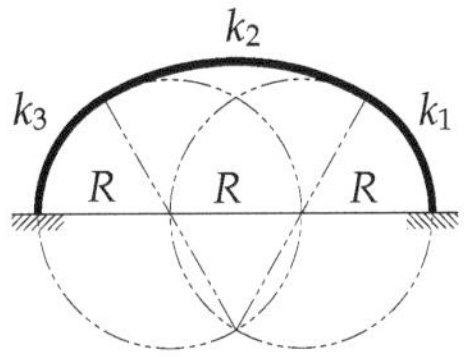

Fig. 11. Curved beam with three subspan of equal span angle of 60°, where k_1=k_3=0.01 and k_2=0.02.

	Present	FEM			CDM
		12 curved elements		24 straight elements	100 Degrees of freedom
Mode		Ref. 1	Ref. 2	Ref. 1	Ref. 1
1	2.683	2.680	2.701	2.685	2.671
2	4.834	4.824	4.828	4.828	4.780
3	9.565	9.536	9.543	9.543	9.413
4	14.585	14.527	14.535	14.535	14.309
5	21.865	21.749	21.751	21.751	21.265

Table 4. Non-dimensional natural frequencies of the curved beam in Fig. 11.

5. Summary

An alternative approach to the dynamic response analysis of one-dimensional distributed parameter systems by applying the concepts of wave motions in elastic waveguides is presented. The analysis techniques are demonstrated using the vibration of multi-span strings, straight beams, and curved beams with general support and boundary conditions, however other one-dimensional systems such as rods and higher order beam models can be treated in the same manner. The present approach allows a systematic formulation that leads to exact, closed-form, distributed transfer functions from which the transient response and frequency response solutions can be obtained. In addition, the present analysis approach results in recursive computational algorithms that always involve computations of fixed-size matrices regardless of the number of subspans, which can be implemented into highly efficient computer codes. Since neither exact nor approximate eigensolutions are required as a priori, the present wave-based approach is suitable for the forced response analysis of non-self-adjoint systems. There are two limiting cases that may affact the analysis accuracy and numerical efficiency of the present wave approach: (1) when the waveguide contains a *very* small amount of inertia or flexibilty (such as massless elements or rigid bodies), which results in making the wavelenght of the constituent waves larger than the span length of the waveguide; and (2) when the waveguide is extremely flexible or its response contains a very sharp impulsive spike (such as the impulse response example shown in Section 2.4), which result in making the wavelength of the constituent waves unrealistically small. In paractice, however, most engineering structures have resonable amounts of inertia, flxibility, and also damping such that these two limiting cases will be rarely encountered.

6. References

Abate, J. & Valkó, P. P. (2004). Multi-Precision Laplace Transform Inversion, *International Journal of Numerical Methods in Engineering*, Vol.60, pp.979-993

Argento, A. & Scott, R. A. (1995). Elastic Wave Propagation in a Timoshenko Beam Spinning about Its Logitudinal Axis, *Wave Motion*, Vol.21, pp. 67-74

Cremer, L.; Heckl, M. & Ungar, E. E. (1973). *Structure-Borne Sound*, Springer, Berlin, Germany

Fahy, F. (1987). *Sound and Structural Vibration*, Academic Press, Orlando, USA

Graff, K. F. (1975). *Wave Motion in Elastic Solids*, Ohio State University Press, Columbus, Ohio, USA

Kang, B. & Tan, C. A. (1998). Elastic Wave Motions in an Axially Strained, Infinitely Long Rotating Timoshenko Shaft, *Journal of Sound and Vibration*, Vol.213, No.3, pp.467-482

Kang, B. (2007). Transfer Functions of One-Dimensional Distributed Parameter Systems by Wave Approach, *ASME Journal of Vibration and Acoustics*, Vol. 129, pp.193-201

Kang, B.; Riedel, R. H. & Tan, C. A. (2003). Free Vibration Analysis of Planar Curved Beams by Wave Propagation, *Journal of Sound and Vibration*, Vol.260, pp.19-44

Krishnan, A. & Suresh, Y. J. (1998) A Simple Cubic Linear Elements for Static and Free Vibration Analyses of Curved Beams, *Computer and Structures*, Vol.68, pp.473-489

Mace, B. R. (1984). Wave Reflection and Transmission in Beams, *Journal of Sound and Vibration*, Vol.97, No.2, pp.237-246

Maurizi, M. J.; Belles, R. E. & Rossi, M. A. (1993). Free Vibration of a Three-Centered Arc Clamped at the Ends, *Journal of Sound and Vibration*, Vol.213, pp.483-510

Mead, D. J. (1994). Wave and Modes in Finite Beams: Application of the Phase-Closure Principle, *Journal of Sound and Vibration*, Vol.171, pp.695-702

Mei, C. & Mace, B. R. (2005). Wave Reflection and Transmission in Timoshenko Beams and Wave Analysis of Timoshenko Beam Structures, *ASME Journal of Vibration and Acoustics*, Vol.127, pp.382-394

Tan, C. A. & Kang, B. (1999). Wave Reflection and Transmission in an Axially Strained, Rotating Timoshenko Shaft, *Journal of Sound and Vibration*, Vol.213, pp.483-510

Yang, B. & Tan, C. A. (1992). Transfer Functions of One-Dimensional Distributed Parameter Systems, *ASME Journal of Applied Mechanics*, Vol. 58, pp.1009-1014

Yang, B. (1996). Closed-Form Transient Response of Distributed Damped Systems, Part I: Modal Analysis and Green's Function Formula, *ASME Journal of Applied Mechanics*, Vol. 63, pp.997-1003

Yang, B. (1996). Closed-Form Transient Response of Distributed Damped Systems, Part II: Energy Formulation fo Constrained and Combined Systems, *ASME Journal of Applied Mechanics*, Vol. 63, pp.1004-1010

Yong, Y. & Lin, Y. K. (1989). Propagation of Decaying Waves in Periodic and Pieccewise Periodic Structures of Finite Length, *Journal of Sound and Vibration*, Vol.129, pp.99-118

2

Phase Diagram Analysis for Predicting Nonlinearities and Transient Responses

Juan Carlos Jáuregui
CIATEQ, A.C.
Mexico

1. Introduction

The developments of new manufacturing processes have impacted modern machinery. Nowadays, mechanical parts are produced with tighter tolerances that allow very high precise assemblies. On the other hand, new materials and design techniques have developed lighter elements. Thus, modern machinery operates at very high speeds and accelerations, which, in many cases, shows nonlinear dynamic behaviors.

Intelligent manufacturing systems require on line monitoring equipment coordinated by the control systems. Traditionally, monitoring systems are based on the Fast Fourier Transform (FFT), which, due to its basis is unable to identify transient responses, and nonlinear behaviors. On the other hand, the FFT requires a considerable processing time that limits its application to early fault detections. The development of new sensors, signal processing techniques and faster microprocessors are key elements for modern monitoring systems. These systems require a better understanding of the machine response and the nature of the output signals.

The evolution of the phase space, or phase diagram, represents how the dynamic system evolves in time. Nonlinearities and transient responses can be determined from the smoothness of the tangent vector of the phase diagram accordingly to Liouville's theorem.

Therefore, the phase diagram is a useful technique for predicting transient and nonlinear behavior in mechanical systems. Even more, the phase diagram can be implemented electronically for on line monitoring, and it can identify faults in real time.

Mathematical modeling refers to the use of mathematical language to simulate the behavior of a system. Its role is to provide a better understanding and characterization of the system. In the theory of mechanical vibrations, mathematical models are helpful for the analysis of dynamic behavior of the structure being modeled (Kerschen et al. 2006). Even with advanced computers, experimental testing and system identification help designers to evaluate numerical predictions with experimental data. The term "system identification" is sometimes used in a broader context and may also refer to the extraction of information about the structural behavior directly from experimental data. In this case, it is referred as any systematic way of deriving models from experimental data. This is the main objective of any machinery monitoring systems (Masri 1994).

For linear systems, modal analysis is the most popular approach for system identification. It can describe the behavior of a system for any input. Examples of this are: Ibrahim time domain method, eigensystem realization algorithm, stochastic subspace identification method, polyreference least-square complex frequency domain method among others.

In structural dynamics, typical sources of nonlinearities are:

- Large displacements, large deformations
- Inertia nonlinearities
- Material nonlinearities
- Dry friction effects
- Boundary conditions

Also, viscoelastic supports show marked nonlinear behavior. And it is quite common to find nonlinearities in a damaged structure. Even though, these are two sources of nonlinearities, viscoelastic supports have a stable response, whereas a damaged structure will developed a drastic change in the system's stiffness that can show jumps and chaos.

As it was stated before, there is a need for designing lighter and more flexible machinery and structures. The basic principles that apply to a linear system, and that are the basis for modal analysis, are no longer valid. These phenomena include jumps, bifurcations, saturation, subharmonics, superharmonics, internal resonances, limit cycles, modal interactions and chaos.

One of the ways to study nonlinear systems is the linearization approach. Weakly nonlinear systems were analyzed using the perturbation theory which includes averaging, Lidsteadt-Poincare technique and the method of multiple scales. There are new methodologies such as nonlinear energy pumping, (Wiercigroch & Pavlovskaia, 2008). In particular, nonlinear normal modes (NNMs) and nonlinear multi-modes (NMMs) have been constructed by using the method of multiple scales. This is to analyze the vibration responses by monitoring the modal responses and mode interactions.

The development of structural models from experimental measurements requires methods such as "nonlinear system identification". There is a significant difference from the linear systems, each nonlinear structure has a unique nature, and thus it is very difficult to have a general method to describe a nonlinear system. Therefore, for every type of nonlinearity a different method is required.

An example of a nonlinear system is the Duffing oscillator; it represents a typical example of a polynomial form of restoring force, whereas hysteric damping is an example of a non-polynomial form of nonlinearity. This represents a major difficulty since there is not a single model (Pai, 2007; Li & Qu, 2007).

In a nonlinear system, its stiffness cannot be easily identified. They proposed to estimate the dynamic behavior from the stochastic response. The period of a free oscillation depends on the energy level. Rüdinger & Krenk (2001) determined the dynamic parameters of a simple nonlinear system (equivalent to a nonlinear oscillator).

An area of major interest is the application of simulation technique to nonlinear system identification. There are a vast number of techniques that have been applied to this topic. Most of them are developed from the identification of uncertain quantities. Monte Carlo simulation is a universal method that can provide accurate solutions for problems involving nonlinearities (Schuëller 1997). The major advantage on Monte Carlo simulation is that accurate solutions can be obtained for any problem whose deterministic solution is known. The disadvantage is that it is time-consuming. The most important part is the generation of sample functions of the stochastic process:

- Stationary or non-stationary
- Homogenous or non-homogeneous
- One dimensional or multidimensional
- Single variable or multivariable

- Gaussian or non-Gaussian

For the nonlinear system identification techniques, there are two broad categories: parametric and non-parametric methods.

Parametric methods assume that the system is represented by a mathematical model. Identification consists on the estimation of the model parameters from the experimental data. ($\ddot{x} + \zeta\dot{x} + \omega^2 x = f(t)$ ω and ζ are estimated). These methods also allow for the design verification.

Nonparametric methods refer to techniques which lack of a mathematical model. They take a "system" approach and fit the input-output relationship. (Examples: Auto-Regressive-Moving-Average, Volterra Wiener-Kerner, etc.) Their limitations are the type of input signals, they required many parameters to find a solution. The model could introduce errors that are not related to the system, and noise measurements could be introduced into the model parameters. This is the main source of uncertainty.

Masri (1994) developed a hybrid approach for the identification of nonlinear systems. He applied a parametric approach for the identification of the linear terms and the well know nonlinear terms, and a parametric approach for describing the unknown nonlinear terms.

The approximation is defined from the equation of motion as:

$$M(\ddot{x}) + C(\dot{x}) + K(x) + f_N(t) = f(t), \tag{1}$$

where $f_N(t)$ includes the nonlinear non-conservative forces

$$f_N(t) = f(t) - M(\ddot{x}) - C(\dot{x}) - K(x). \tag{2}$$

The right hand side can be determined from a parametric modeling and $f(t)$ is a well known input function.

$f_N(t)$ can be modeled as a combination of parametric and non-parametric term, this is what Masri (1994) described as a hybrid model. He approximated the $f_N(t)$ as vector h were each element $h_i(t)$ is a function of the acceleration, velocity and position vectors associated with each degree of freedom. Masri (1994) showed that the nonlinear terms can be visualized in the phase diagram and they can be isolated by subtracting the linear components from the measured data.

The development of the nonlinear dynamic theory brought new methods for recognition and prediction of nonlinear dynamic response (Yang 2007).

The nonlinear dynamic and chaos theory can be used to describe the irregular, broadband signals, which are generic in non-linear dynamical systems, and extracting some physically interesting and useful features from such signals. Fractal dimension, such as the capacity dimension, correlation dimension, and information dimension, developed by the Nonlinear dynamic and chaos theory, is a promising new tool to interpret observations of physical systems where the time trace of the measured quantities is irregular. The phase diagram and Poincare maps of chaotic systems have a fractal structure. We can recognize, classify and understand such maps of chaos by measuring the stability of the phase diagram.

Vela et al. (2010) applied a detrended fluctuation analysis (DFA), adapted for time-frequency domain, to monitor the evolution nonlinear dynamics. The underlying idea behind the application was to use the Hurst exponent, an index of the signal fractal roughness, to detect dominance of unstable oscillatory components in the complex, presumably stochastic, dynamics of machine acceleration. In early stages of machinery faults the signal-noise ratio is very low due to relatively weak energy signals. Other authors have studied the effect of a weak periodic signal in the chaotic response of a nonlinear oscillator (Li & Qu, 2007; Modarres et al. 2011). Liu (2005) developed a visualization method for nonlinear chaotic systems.

One of the advantages of the display identification is the representation of the phase diagram as a three-dimensional plot. In this way the phase diagram can be related to the frequency and the dynamic identification of the system. According to Taken's theorem, a dynamic system can be obtained by reconstructing the phase diagram (Wang, G. et al. 2009; Wang, Z., et al. 2011; Ghafari et al. 2010).

Karpenko et al., (2006) applied the phase diagram in the identification of nonlinear behavior of rotors. They also demonstrated that rubbing is nonlinear and can be identified as a chaotic system. Mevela &. Guyade (2008) developed a model for predicting bearing failures.

In this chapter, the application of the phase space, or phase diagram, to the identification of nonlinearities and transient function is presented. The theoretical background is discussed in next section, and afterwards its application to the most relevant mechanical systems is presented.

2. Phase diagram definition

The analysis and modeling of dynamic systems can be done from a Lagrangian approach or from a Hamiltonian approach. The Lagrangian approach describes how position and velocity change in time. The Hamiltonian approach describes how position and momentum change in time. The position and momentum of a particle specifies a point in a space called the "phase space", "phase plane", "phase diagram", among others (Nichols, 2003).

A particle traces out a path in a space R^n

$$q: \mathcal{R} \to \mathcal{R}^n \tag{3}$$

where R represents time domain, R^n represents the space domain and q represents the position of a particle at an instant t.

From Newton's law

$$F(t) = \frac{d}{dt}\left(m\dot{q}(t)\right) = \frac{d}{dt}(p(t))\,, \tag{4}$$

with the restriction that $F(t)$ is a smooth function.

The potential energy of a multi-particle system will have the form

$$V_i(t) = \sum_{i \neq j} V_{ij}\left|q_i(t) - q_j(t)\right| \tag{5}$$

where

$$V_{ij} = -\int_c^r f_{ij}(s)ds \tag{6}$$

and f_{ij} is the force acting between particle i and j.

Hamilton's principle is defined as:

$$H(q,p) = \frac{p^2}{2m} + V(q) \tag{7}$$

and

$$\frac{\partial H(q,p)}{\partial p_i} = \frac{p_i}{m} \tag{8}$$

$$\frac{\partial H(q,p)}{\partial q_i} = \frac{\partial V(q)}{\partial q}. \tag{9}$$

Thus

$$\frac{d}{dt}q_i(t) = \frac{\partial H}{\partial p_i}(q(t),p(t)) \tag{10}$$

$$\frac{d}{dt}p_i(t) = -\frac{\partial H}{\partial q_i}(q(t),p(t)),$$

where the dyad $(q(t),p(t))$ represents the phase space of a particle, and $(q,p) \in \mathcal{R}^n x \mathcal{R}^n$. If the phase space can be represented as a smooth function $\varphi: \mathcal{R}^n x \mathcal{R}^n \to \mathcal{R}$, then it represents the system's evolution in time. Thus, for a system with n particles

$$\frac{d\varphi}{dt} = \sum_i \frac{\partial \varphi}{\partial q_i}\frac{dq_i}{dt} + \frac{\partial \varphi}{\partial p_i}\frac{dp_i}{dt}. \tag{11}$$

Using Hamilton's equation

$$\frac{d\varphi}{dt} = \sum_i \frac{\partial \varphi}{\partial q_i}\frac{\partial H}{\partial p_i} - \frac{\partial \varphi}{\partial p_i}\frac{\partial H}{\partial q_i} \tag{12}$$

For example, a simple harmonic is represented as

$$m\ddot{q}(t) + kq(t) = 0 \tag{13}$$

with its well known solution

$$q(t) = Asin(\omega t) + Bcos(\omega t) \tag{14}$$

$$p(t) = m(A\omega cos(\omega t) - B\omega sin(\omega t)) \tag{15}$$

where

$$A = q(0)$$

$$(16)$$

$$B = \frac{p(0)}{m}.$$

The Hamiltonian can be written as:

$$H(q,p) = \frac{1}{2}\left(\frac{p^2}{m} + kq^2\right). \tag{17}$$

The field vector operator is defined as:

$$v_H = p\frac{\partial}{\partial q} - q\frac{\partial}{\partial p} \tag{18}$$

and the flow field is found as:

$$\varphi_t = [q(t),p(t)]. \tag{19}$$

In this case

$$\varphi_t = \left[q(0)\sin(\omega t) + \frac{p(0)}{m}\cos(\omega t)\,,\, q(0)m\omega\cos(\omega t) - p(0)\omega\sin(\omega t)\right] \tag{20}$$

This flow field represents an ellipse at any time t.
The dynamic stability is determined from Liouville's theorem, (the phase space volume occupied by a collection of systems evolving according to Hamilton's equations of motion will be preserved in time):

$$\frac{dH}{dt} = \frac{\partial H}{\partial t} + \sum_{i=1}^{n}\left(\frac{\partial H}{\partial q_i}\dot{q}_i + \frac{\partial H}{\partial p_i}\dot{p}_i\right) = 0 \tag{21}$$

It can be shown that

$$\sum_{i=1}^{n}\left(\frac{\partial^2 H}{\partial q_i \partial p_i} - \frac{\partial^2 H}{\partial q_i \partial p_i}\right) = 0 \tag{22}$$

This conservation law states that a phase diagram volume will be preserved in time; this is the statement of Liouville's theorem.

3. Application to nonlinear mechanical systems

3.1 Gears

As a complete system a gear box contains gears or teeth wheels, shafts, bearings, rolling bearings, lubrication pumps, tubes, valves and other devices such as heat exchangers. Therefore, all these individual elements have gone through a development process by themselves, but as an integrated system they have challenged engineers with highly interesting problems. The one of particular interest is gear vibrations, which is always undesirable, and also generates noise. The dominant cause of gear noise is the Transmission Error; it is the deviation from a perfect motion between the driver and the driven gears. And it is the combination of different gear variations, such as non perfect tooth profiles, pitch errors, elastic deformations, backlash, etc. The simplest type of noise is a steady note which may have a harmonic content at the gear mesh frequency. This frequency is normally modulated by the rotating frequency. Modulated noise is often described as a buzzing sound. In general, gears show a frequency modulated spectrum with a distinguishable mesh frequency and side bands spaced at the shaft rotating frequency. Other noises are associated with pitch errors. They are described as scrunching, grating, grouching, etc. They contain a wide range of frequencies that are a lot higher than the rotating frequency. White noise can also be present and it may be associated with loss of contact between the teeth. (Jauregui & Gonzalez, 2009).

Gear box vibration is a typical nonlinear vibration phenomenon. Its nonlinear behavior comes from the discontinuities in the stiffness of the system, which comes from the combination of two teeth acting in conjunction. Thus, the stiffness of a gear pair varies with the angular position, except in very specific gear designs. One of the main features of gear pair stiffness is that it changes drastically as a function of the number of teeth in simultaneous contact. Ideally, a pair of gears transmits motion at a constant speed.

In most gear pair systems, torsional motion is coupled by the gear pair stiffness; therefore a two degree of freedom model will reflect accurately most practical applications. If it is necessary to include other effects, increasing the degrees of freedom could accommodate other compliances that are present in the system.

Many researchers and engineers have developed a significant number of gear dynamic models. Most of them have been developed for the prediction of noise and vibrations, and they have demonstrated that gear vibrations are highly nonlinear. In this chapter we present one of the most commonly used model that is widely accepted. It was demonstrated that a simplified lumped-mass model is adequate for small transmissions. (Chang 2010).

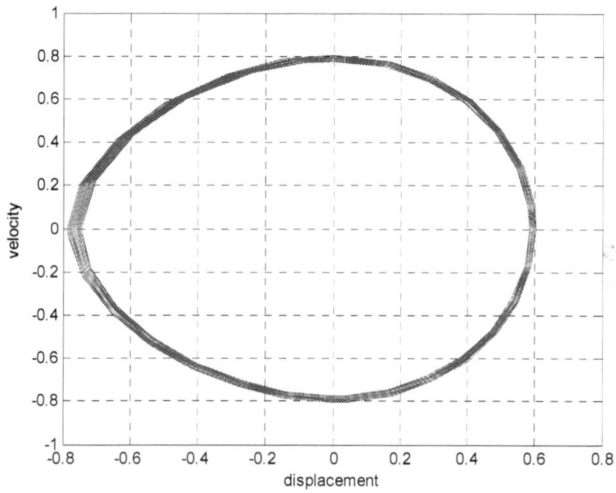

Fig. 1. Phase diagram of a pair of gears under free vibration.

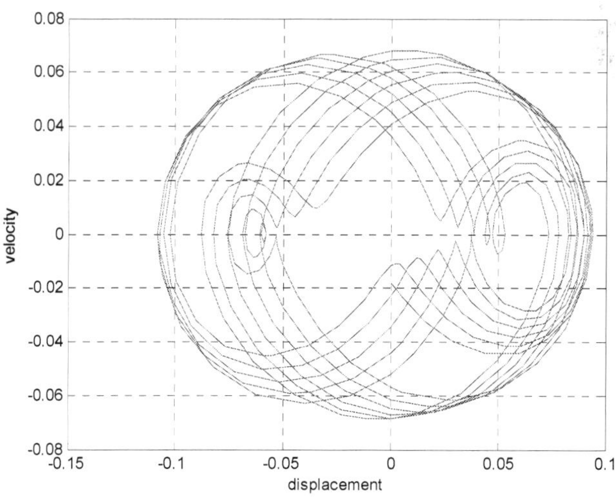

Fig. 2. Phase diagram of a pair of gears with an external excitation of 0.4 of the linear natural frequency

In this case, it is important to identify the effect of the nonlinear gear action in a phase diagram. From a simple lumped mass model, it is sufficient to identify the nonlinear response of a transmission. Fig 1 represents the phase diagram of the free vibration response. In this case, a small damping coefficient was included in the model. It is noticeable how the nonlinear stiffness deforms the phase space pattern, and instead of producing an ellipse, it forms a lemon shape. For practical purposes, this pattern is stable at any time.
Fig. 2, represents the forced vibration response with an external excitation at 0.4 ω_N. It is clear to see how the stable pattern disappears, and two attracting poles are formed around the origin of the phase space. This behavior is similar to a nonlinear Duffing oscillator. Fig. 3 shows the same system but with an external excitation beyond its first linear natural frequency. In this case, the instability is larger and number of attracting poles increases and the velocity amplitude almost doubles the other two cases.
Gears have a characteristic phase diagram; it changes from a stable non-elliptical pattern to a chaotic phase space. This drastic change is quite significant and, with an appropriate monitoring system, it can detect early faults in the gear teeth, or damaging effects caused by changes in the operating conditions.

3.2 Discontinuous stiffness

Stiffness discontinuities are present in many mechanical systems. It is one reason why gears have a nonlinear dynamic behavior. Another type of stiffness discontinuity is found in cracked structures. Andreaus & Baragatti (2011) demonstrated that a cracked beam behaves as a discontinuous stiffness system. This discontinuity is a function of the beam's displacement, thus the stiffness is lower when the beam's movement opens the crack and the stiffness increases when the movement closes the crack. Also large deformations can produce a similar pattern as a system with stiffness discontinuities, (Machado et al. 2009), (Mazzillia et al.,2008).

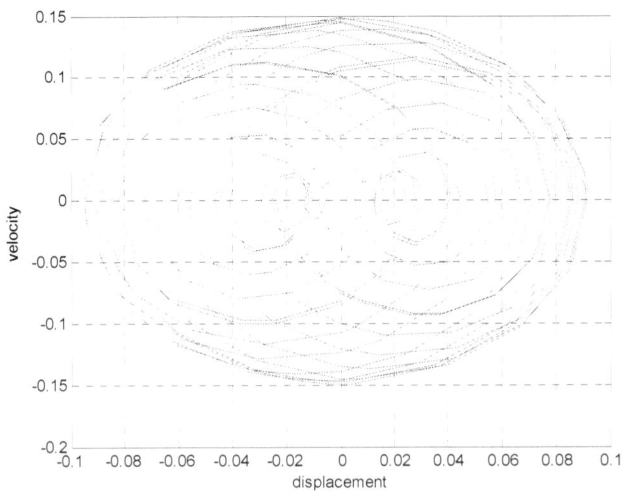

Fig. 3. Phase diagram of a pair of gears with an external excitation of 1.6 of the linear natural frequency

A typical pattern of a beam under large deformations can be seen in Fig. 5 (Jauregui & Gonzalez, 2009b). The elliptic shape evolves into a rectangular shape with two attracting poles.

This behavior is found in very large and thin structures such as wind turbine blades or helicopter blades. The stability of these structures depends entirely on internal damping capabilities.

3.3 Bearings

Most of the dynamic models of rolling bearings consider that their stiffness is a function of the frequency and the displacement. This characteristic makes its dynamic behavior different from other mechanical elements. And, as was stated in the introduction, it is quite complicated to establish a single nonlinear mode of vibration. Thus, in a bearing system, strange motions appear due to the nature of the stiffness function. To describe these strange motions, tools specific to chaotic dynamics have to be introduced. Fourier spectra are convenient for detecting sub- or super-harmonics of a component, also in the case of complete chaotic behavior, but the quasi-periodic motion is impossible to detect except for the ideal case of two components. Some recent studies have used phase diagrams and Poincare´ sections. An extremely efficient technique is then to sample the phase diagram points using a convenient clock frequency, in order to obtain a limited number of points. The resulting shape is an excellent tool to characterize sub-harmonic, quasi-periodic or chaotic motions.

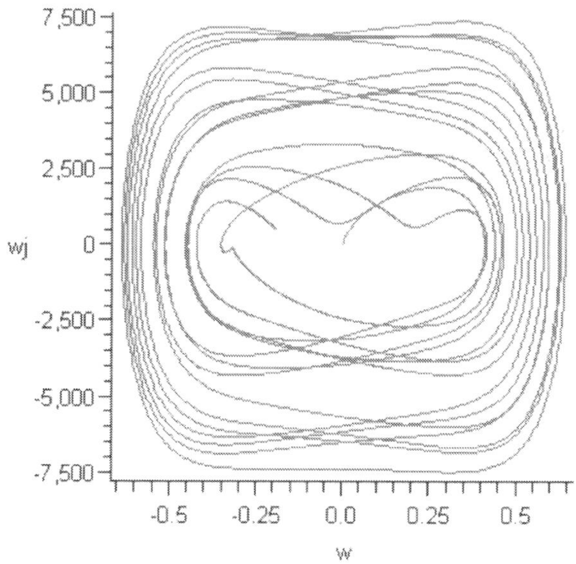

Fig. 5. Phase diagram of a beam under large deformations

A typical ball-bearing system consists of five contact parts: the shaft, the inner ring, the rolling elements, outer ring and the housing. The deformation of each part will influence the load distribution and, in turn, the service life of the bearing. It is well known that classical calculation methods cannot predict accurately load distributions inside the bearing. Ball bearings (Fig 6) are very stiff compared with sliding bearings; their stiffness can be approximating as a set of individual springs; where the number of springs supporting the shaft varies with the angular position of the shaft. This variation depends upon the kinematics of the ball roller as it moves around the shaft. Thus, the ratio of rotation of the ball as a function of shaft's rotation is determined as

$$\omega_b = \frac{D\omega_s}{d} \tag{24}$$

The fundamental principle of a rolling bearing is that the ball or roller translates around the shaft, eliminating must of the friction; then the ball's angular translation is found as (D is the pitch diameter and d is the roller diameter)

$$\phi = \frac{d\cos(\omega_b t)}{D + \frac{d}{2}} \cos(\alpha) \tag{25}$$

The number of balls, or rolls in contact are determined from Fig. 7. The nonlinear characteristic of the rolling bearing is the ball-track deformation. The ball-track stiffness is calculated with the Hertz equation. Since the balls translate around the shaft, the number of balls supporting the load varies with the angular position of the shaft; this translation effect modifies the overall stiffness of the bearing. Although this variation may be small, it creates a nonlinear vibration, which turns out to be relatively difficult to identify in field problems.

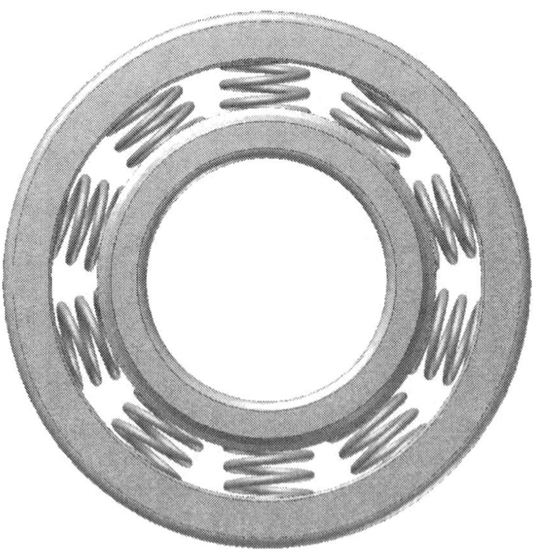

Fig. 6. Schematic representation of a roller bearing

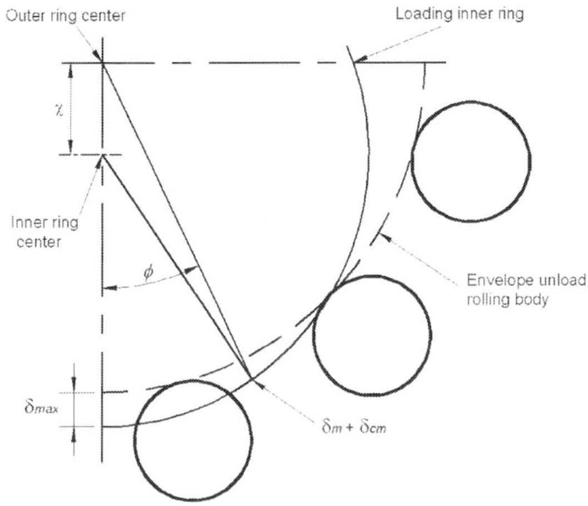

Fig. 7. Radial displacement of a shaft mounted on ball bearings

Rolling bearings generate transient vibrations due to stiffness nonlinearities and structural defects. There are four external sources of vibration; two of them are associated with the angular velocity of the ball ω_b and their angular translation ω_ϕ. The other two frequencies are related to structural defects on the inner and outer tracks. These external frequencies excite the nonlinear terms. The stiffness of the ball as a function of the deformation is almost constant:

$$P_i = E\sqrt{\frac{Dd}{(D-d)}\left(\frac{\delta_i}{\alpha_H}\right)^3} \tag{26}$$

The nonlinear effect comes from the combination of balls deformation as they roll around the shaft. The rolling bearing can be modeled as a mass-spring system.

$$\delta_i = \delta_{\max}\cos\left[\frac{\pi}{2} - i\left(\phi + \frac{2(i-1)\pi}{N}\right)\right] \tag{27}$$

The spring stiffness is determined from Fig. 8. Similarly as the gear mesh stiffness, rolling bearings exhibits a periodic function, thus it can be expanded as a Taylor series:

$$k_x = a_0 + a_1\phi + a_2\phi^2 + a_3\phi^3 \tag{28}$$

Coefficients a_i are function of the number of balls under load, and ϕ represents roller translation angle.

The solution of the dynamic model requires the definition of the transmitted force. Ideally, it should be constant, and equal to the radial force. But, it is not the case; first of all, the radial force varies according to every application, and the rolling bearing itself produces a specific type of excitation forces. These forces are associated with physical defects on the bearing, and there are basically four types of excitation.

One of the challenges of a monitoring system is the identification of early faults in rolling bearings. Failures in bearings start at surface level; thus, they generate a relatively small energy vibration compared to other sources, and its identification is very cumbersome. With the application of phase diagram plots, early failures can be predicted in real time. The process is as follows:

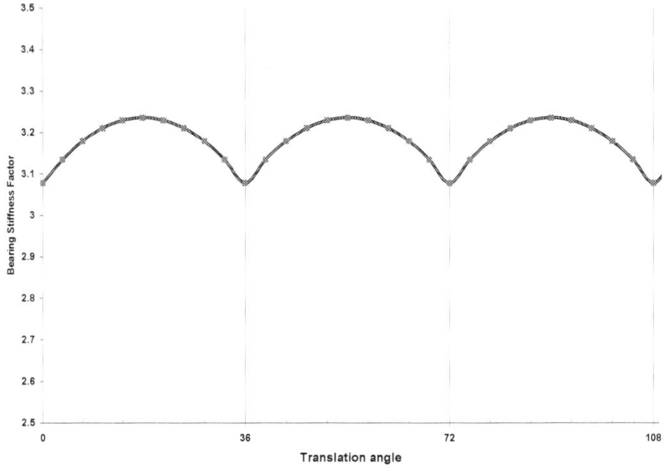

Fig. 8. Bearing stiffness function

Vibrations are measured with a transducer, preferably an accelerometer. Then, the signal is analogically integrated in real time, and the phase diagram is plotted. When the bearing is new, the first diagram (Fig. 9) corresponds to the healthy reference. Since we know that bearings have a nonlinear response, and that this response is the result of its stiffness dependency on frequency, we can monitor the phase diagram in order to "see" the instant when instabilities occur. In this way, if we permanently monitor the "shape" of the phase diagram, and we detect the appearance of instabilities, then we will be able to detect early faults. Fig.9 shows a phase diagram of a healthy bearing. In this figure, we can see four major loops, they correspond to the main frequencies, the unbalance load produces the external loop, and the other three are the mayor bearing frequencies. This diagram shows similar shapes at different time steps.

Fig. 10 shows the phase diagram of a damage bearing. Comparing both diagrams, it is clearly seen that bearing looses stability when there is a defect. This stability change can be detected with an appropriate electronic monitoring system.

3.4 Friction

Dry friction is an important source of mechanical damping in many physical systems. The viscous-like damping property suggest that many mechanical designs can be improved by configuring frictional interfaces in ways that allow normal forces to vary with displacement. The system is positively damped at all times and is clearly stable (Anderson & Ferri 1990) (Oden & Martins, 1985).

Distinctions between coefficients of static and kinetic friction have been mentioned in the friction literature for centuries. Euler developed a mechanical model to explain the origins of frictional resistance. He arrived at the conclusion that friction during sliding motion should be smaller. The experiment proposed by Euler involved the sliding of a body down an inclined plane at slopes close to the critical slope at which sliding initiates. This, of course, would mean that, as soon as sliding initiates, a drop of friction force occurs, the difference between static and kinetic friction forces being responsible for the acceleration of the body down the inclined plane (stick-slip in sliding systems).

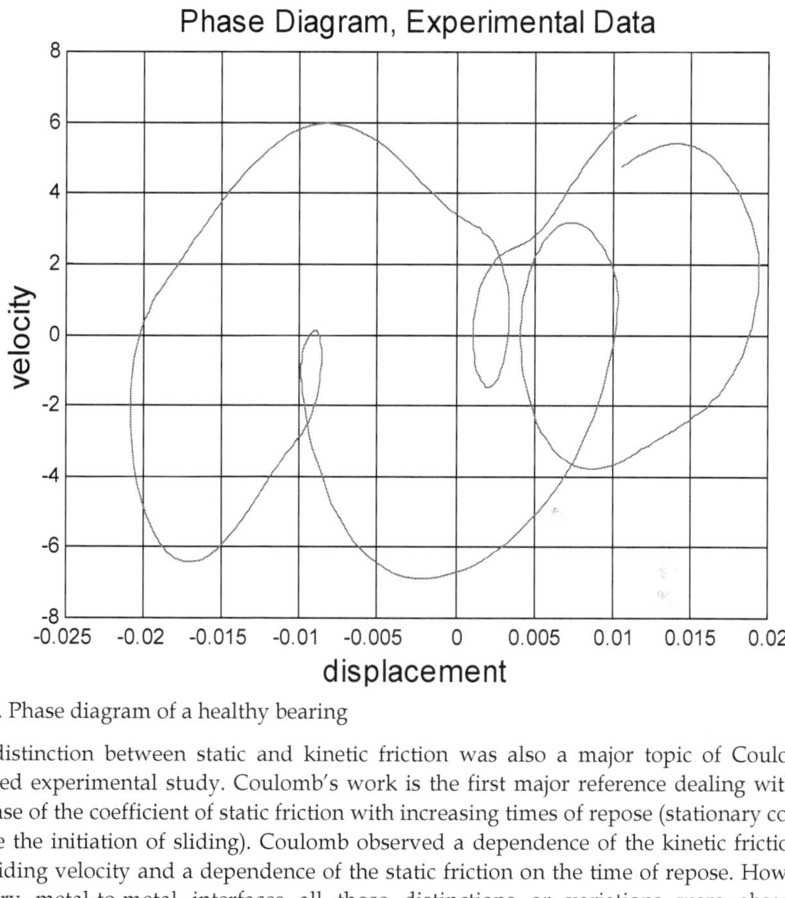

Fig. 9. Phase diagram of a healthy bearing

The distinction between static and kinetic friction was also a major topic of Coulomb's detailed experimental study. Coulomb's work is the first major reference dealing with the increase of the coefficient of static friction with increasing times of repose (stationary contact before the initiation of sliding). Coulomb observed a dependence of the kinetic friction on the sliding velocity and a dependence of the static friction on the time of repose. However, for dry metal-to-metal interfaces all those distinctions or variations were absent or negligible.

In general, the coefficient of kinetic friction would be small and increasing with sliding velocity at low velocities. Then, at some velocity, it would achieve a maximum value after which it would decrease with the increase of speed. The sliding process is not a continuous one; the motion proceeds by jerks. The metallic surfaces "stick" together until, as a result of the gradually increasing pull, there is a sudden break with a consequent very rapid "slip". The surfaces stick again and the process is repeated indefinitely. When the surfaces are of the same metal, the behavior is somewhat different. Large fluctuations in the friction still occur but they are comparatively slow and very irregular. The average value of the frictional force is considerably higher than that found for dissimilar metals and a well-marked and characteristic track is formed during the sliding.

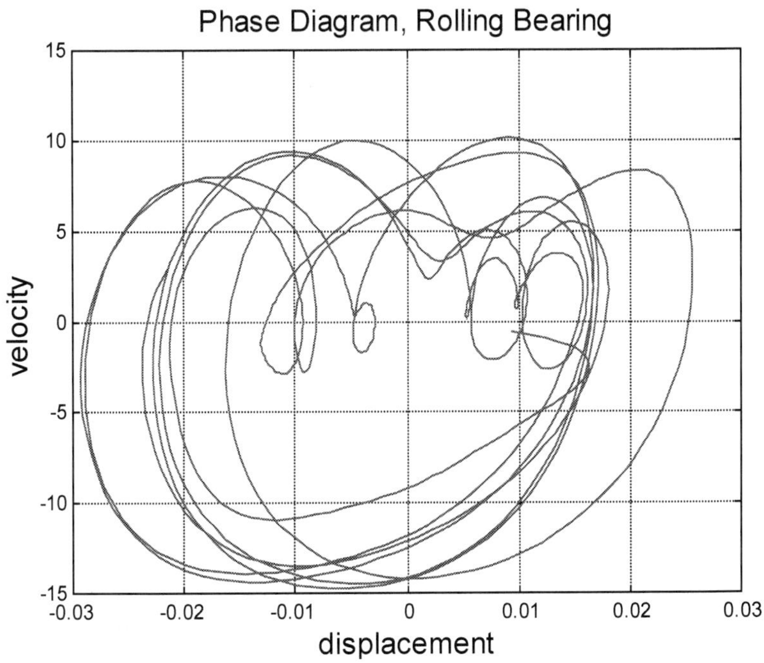

Fig. 10. Phase diagram of a damaged bearing.

It has also been observed that the frequency of the stick-slip motion increases with the increase of the driving velocity and that the maximum value of this frequency approaches the undamped natural frequency of the system, although in some cases the oscillation stops at a level well below that of the natural frequency.

The Martin model (Oden &Martins 1985) considers a two-degree-of-freedom system, where the normal force between the sliding block is its weight, and it is free to separate from the sliding surface upwards

$$F_{\mu} = \mu(y^2)sgn(\dot{x} - v)\ . \tag{29}$$

The dynamic behavior of a single-degree-of-freedom system with amplitude and rate dependant friction forces is presented. A system with amplitude-dependant friction is more likely to experience intermittent sticking. If the system sticks a significant amount of time, the energy dissipation capability may be degraded. Hence, special care is taken in this analysis to examine sticking conditions (in the case of gear teeth action sticky occurs only for very high contact stresses). In general sticking can occur only when the sliding velocity is zero.

The extended friction law is (Anderson &Ferri 1990):

$$F_{\mu} = \mu(C_0 + C_1|x| + C_2|\dot{x}|)sgn(\dot{x}) \tag{30}$$

where x represents the sliding displacement, $\dot{x}$ represents the sliding velocity, C_0 is the normal force, C_1 is the friction interface amplitude, C_2 is the friction interface velocity and μ is the coefficient of friction (in general is the equivalent to the static coefficient of friction).

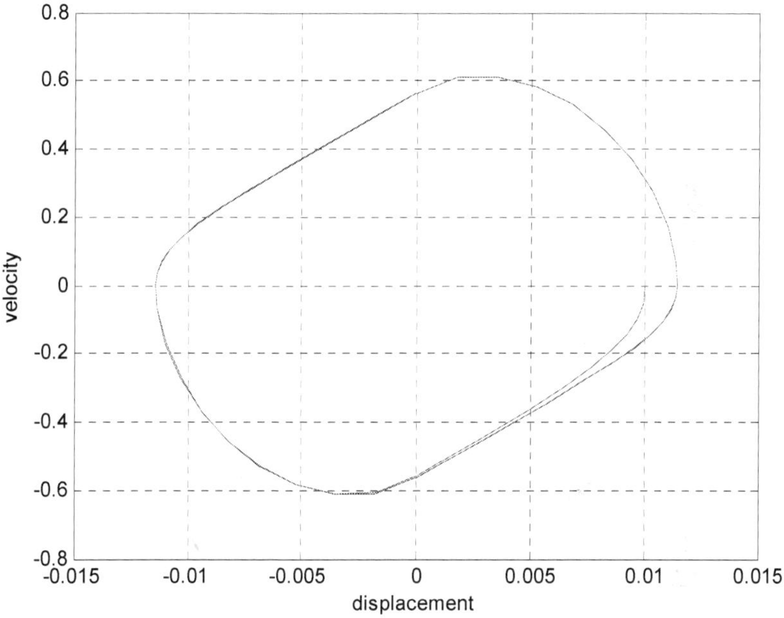

Fig. 11. Phase diagram for a single degree freedom model. Free vibration

The dynamic model for a single-degree-of-freedom system is represented as:

$$m\ddot{x} + c\dot{x} + kx + F_{\mu} = F_e\cos(\omega t). \tag{31}$$

The system is positively damped at all times and it is clearly stable in the sense of Lyapunov. However the system is not asymptotically stable for $C_0 \neq 0$. This condition is identified from the phase diagram when $F_e = 0$ and the initial x=0.1.
The friction force has a particular behavior; it can be observed in Fig. 12.

3.5 Rubbing

One of the most interesting and practically important dynamic responses of rotor systems are caused by bearing clearances, which are mainly due to piecewise nature of stiffness characteristics. It is well known that dynamic interactions in such systems can lead to much more complex nonlinear behavior than in systems with smooth nonlinearities, including existence of grazing bifurcations and untypical routes to chaos such as blowout. In rotor

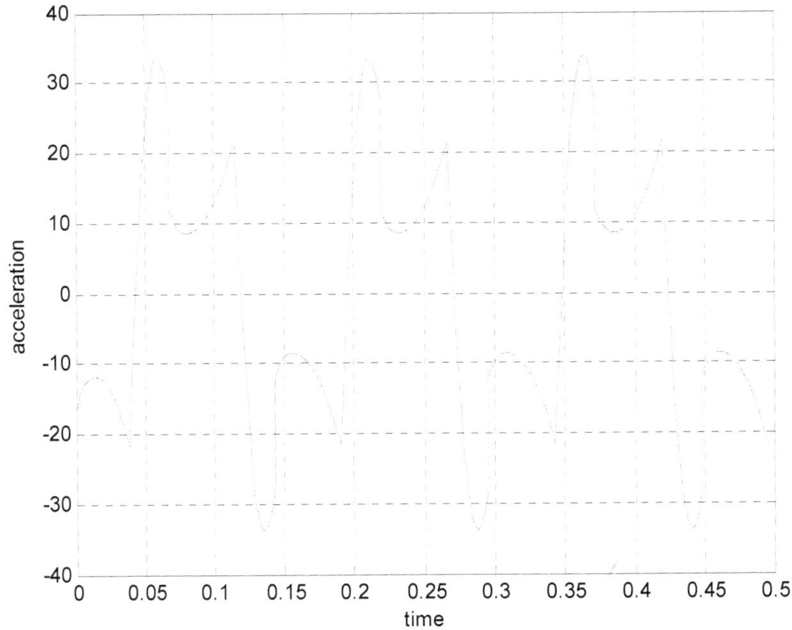

Fig. 12. Friction force produced by the general friction law

systems, such phenomena are caused by intermittent contacts between the components of the rotor system, which can lead to catastrophic failures. Therefore, it is vastly important to conduct experimental verifications in order establish credible mathematical models predicting complex dynamic responses of rotor systems.

There is a big difference between friction and rubbing. Whereas friction has a smooth phase diagram, rubbing develops a chaotic behavior. This chaotic behavior is associated with a sever stiffness discontinuity (Jauregui & Gonzalez 2009a; Rubio & Jauregui 2011). Pure friction acts as a damper and stabilizes the system, whereas, rubbing modifies the stiffness of the system. The stability pattern will depend on the hardness of the surface. If a rotor rubs a hard surface, the phase diagram will show a pattern similar to Fig. 13. If the rotor rubs a soft surface, the phase diagram will have a different pattern (Fig. 14).

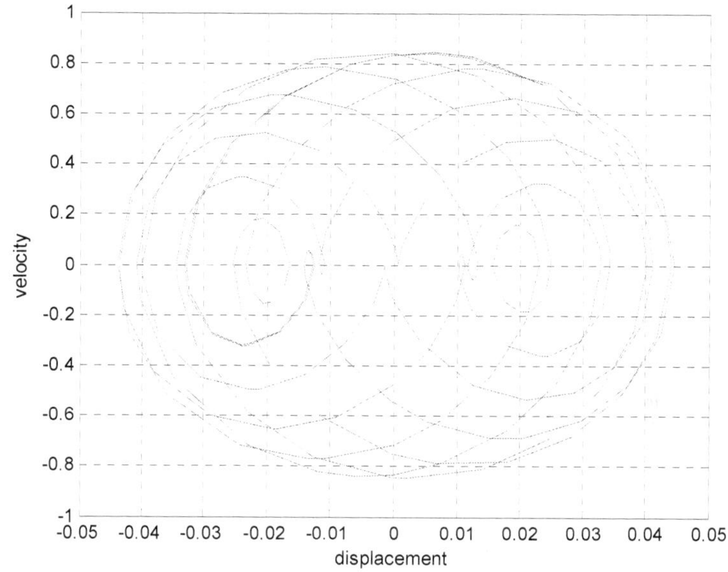

Fig. 13. Phase diagram for a rotor rubbing a hard surface

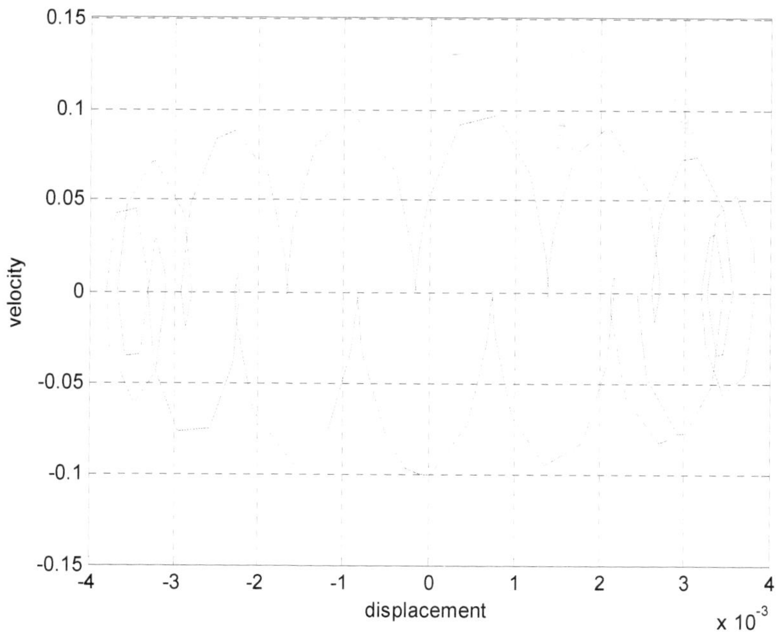

Fig. 14. Phase diagram for a rotor rubbing a soft surface

4. Conclusions

The phase diagram or phase space is a geometric representation of the Hamiltonian function of a dynamic system. It represents the relationship between the potential and kinetic energies at any time. Since the dynamic instability of a system will modify the trajectory of the phase space, the phase diagram will reflect any significant change in the Hamiltonian function. Thus, the phase diagram is a useful tool for predicting and monitoring nonlinear systems, and also it can identify transient responses. Linear systems have stable, well defined elliptical functions.

The phase diagram can be implemented into an electronic device, and it is possible to construct it in real time. The difficulty arises from its interpretation. Therefore, it is important to understand the phase diagrams of each system. It is known that there is no single model that can describe a nonlinear system. Some mechanical elements can be represented with certain models, but their dynamic response will depend not only on the model, but also in the system's sensibility to the nonlinear terms.

The phase diagram can be used to determine the system response and it can identify when a dynamic system becomes instable. For most nonlinear systems, the phase space changes from a smooth simple geometry, to a completely different pattern. Even more, under stable conditions, the phase space repeats its shape periodically, whereas under an unstable condition, the pattern changes as a function of time.

5. References

Kerschen, G., Worden, K., Vakakis, A. & Golinvala, J. (2006). Past, present and future of nonlinear system identification in structural dynamics, *Mechanical Systems and Signal Processing*, Vol. 20, pp 505–592

Masri, S., (1994). A hybrid parametric/nonparametric approach for the identification of nonlinear systems, *Probabilistic Engineering Mechanics*, Vol. 9, pp 47-57

Wiercigroch, M. & Pavlovskaia, E., (2008) Non-linear dynamics of engineering systems, *International Journal of Non-Linear Mechanics*, Vol. 43 pp 459 -461

Pai, F., (2007) Nonlinear vibration characterization by signal decomposition, *Journal of Sound and Vibration*, Vol. 307 pp 527–544

Yang, J., Zhang, Y. & Zhu, Y. (2007). Intelligent fault diagnosis of rolling element bearing based on SVMs and fractal dimension, *Mechanical Systems and Signal Processing* Vol. 21 pp 2012–2024

Li, C. & Qu, L. (2007) Applications of chaotic oscillator in machinery fault diagnosis, *Mechanical Systems and Signal Processing* Vol. 21 pp 257–269

Rüdinger, F. & Krenk, S., Non-parametric system identification from non-linear stochastic response, *Probabilistic Engineering Mechanics*, Vol. 16 pp 233-243

Schuëller, G. (1997) A State-of-the-Art Report on Computational Stochastic Mechanics, *Probabilistic Engineering Mechanics*, Vol. 12, No. 4, pp. 197-321

Vela, L., Jauregui, J., Rodriguez, E., & Alvarez, J. (2010) Using detrended fluctuation analysis to monitor chattering in cutter tool machines, *International Journal of Machine Tools & Manufacture* Vol. 50 pp 651–657

Modarres,Y., Chasparis,F., Triantafyllou, M., Tognarelli ,M. & Beynet, P. (2011) Chaotic response is a generic feature of vortex-induced vibrations of flexible risers, *Journal of Sound and Vibrations,* Article in press.

Liu, B., (2005) Selection of wavelet packet basis for rotating machinery fault diagnosis, *Journal of Sound and Vibration* Vol. 284 pp 567–582

Wang, G., Li, Y., & Luo, Z., (2009) Fault classification of rolling bearing based on reconstructed phase space and Gaussian mixture model, *Journal of Sound and Vibration* Vol.323 pp 1077–1089

Wang, Z., Akhtar, I., Borggaard, J., & Iliescu, T., (2011) Two-level discretizations of nonlinear closure models for proper orthogonal decomposition, *Journal of Computational Physics* Vol.230 pp 126–146

Ghafari, S., Abdel, E., Golnaraghi, F. & Ismail, F., (2010) Vibrations of balanced fault-free ball bearings, *Journal of Sound and Vibration* Vol. 329 pp 1332–1347

Karpenko, E., Wiercigroch, M., Pavlovskaia, E. & Neilson, R., (2006). Experimental verification of Jeffcott rotor model with preloaded snubber ring, *Journal of Sound and Vibration* Vol. 298 pp 907–917

Mevela,B. & Guyaderb, J. (2008) Experiments on routes to chaos in ball bearings, *Journal of Sound and Vibration* Vol. 318 pp 549–564

Nichols, J. (2003), Structural health monitoring of offshore structures using ambient excitation, *Applied Ocean Research* Vol. 25 pp 101–114

Jauregui, J. & Gonzalez, O., (2009a), *Mechanical Vibrations of Discontinuous Systems* (1st edition), Nova Science Publishers, ISBN: 978-1-60876-126-5, New York

Jauregui, J. & Gonzalez, O., (2009b),Non-linear vibrations of slender elements, In: *Mechanical Vibrations measurements, effects and control,* Sapri, R., pp 557-588, Nova Science Publishers, ISBN: 978-1-60692-036-7, New York

Chang, C., Strong nonlinearity analysis for gear-bearing system under nonlinear suspension -bifurcation and chaos, *Nonlinear Analysis: Real World Applications* Vol.11 pp 1760–1774

Andreaus, U. & Baragatti, P., (2011) Cracked beam identification by numerically analysing the nonlinear behaviour of the harmonically forced response, *Journal of Sound and Vibration* Vol. 330 pp 721–742

Machado, L., Lagoudas, D. & Savi, M., (2009) Lyapunov exponents estimation for hysteretic systems, *International Journal of Solids and Structures* Vol. 46 pp 1269–1286

Mazzillia, C., Sanches, C., Netoa, B., Wiercigrochb, M. & Keber, M., Non-linear modal analysis for beams subjected to axial loads: Analytical and finite-element solutions, *International Journal of Non-Linear Mechanics* Vol. 43 pp 551 - 561

Anderson, J., &. Ferri, A., (1990) Behavior of a single-degree-of-freedom-system with a generalized friction law, *Journal of Sound and Vibration* Vol. 140(2), pp 287-304

Oden, J., Martins, J., (1985) Models and computational methods for dynamic friction phenomena, *Computer methods in applied mechanics and engineering,* Vol. 52, pp 527-634

Rubio, E., Jauregui, J., (2011), Time-Frequency Analysis for Rotor-Rubbing Diagnosis, *Advances in Vibration Analysis Research,* Ebrahimi, F., ISBN 978-953-307-209-8 , InTech Publishers

3

A Levy Type Solution for Free Vibration Analysis of a Nano-Plate Considering the Small Scale Effect

E. Jomehzadeh[1,2] and A. R. Saidi[1]
[1]*Department of Mechanical Engineering, Young Researchers Club, Kerman Branch, Islamic Azad University, Kerman*
[2]*Shahid Bahonar University of Kerman, Young Researchers Society, Kerman, Iran*

1. Introduction

Experimental results show that as length scales of a material are reduced, the influences of long-range interatomic and intermolecular cohesive forces on the mechanical properties become prominent and cannot be neglected. It is well known that surfaces and interfaces in nano structures behave differently from their bulk counterparts. For nanostructures with size less than 100nm, the surface to volume ratio is significant and the effective properties are altered by surface and nonlocal effects. Therefore, at nanolength scales, size effects often become prominent, the causes of which need to be explicitly addressed especially with an increasing interest in the general area of nanotechnology (Sharma et al., 2003).

Due to the vast computational expenses of nano-structures analyses when using atomic lattice dynamics and molecular dynamic simulations, there is a great interest in applying continuum mechanics for analysis of nano-structures. Classical continuum elasticity, which is a scale free theory, cannot predict the size effects. Nonlocal continuum mechanics allows one to account for the small length scale effect that becomes significant when dealing with microstructures or nanostructures. It has been showed that it is possible to represent the integral constitutive relations of nano-structures in an equivalent differential form (Eringen, 1983). Eringen presented a nonlocal elasticity theory to account for the small scale effect by specifying the stress at a reference point is a functional of the strain field at every point in the body. Since then, many studies have been carried out nonlocal theory of elasticity for bending, buckling and vibration analyses of nano-structures.

Small scale effect on static deformation of micro- and nano-rods or tubes is revealed through nonlocal Euler–Bernoulli and Timoshenko beam theories by Wang and Liew (2007). Li and Wang (2009) investigated a theoretical treatment of Timoshenko beams, in which the influences of shear deformation, rotary inertia, and scale coefficient are taken into account.

Murmu and Pradhan (2009a) studied vibration response of nano cantilever considering non-uniformity in the cross sections using nonlocal elasticity theory.

Although graphite sheet has many superior properties, such as low electrical and thermal conductivities normal to the sheet but high electrical and thermal conductivities in the plane of the sheet, relatively little research have been reported in the literature for mechanical analyses of graphene sheets.

Kitipornchai et al. (2005) used the continuum plate model for mechanical analysis of graphene sheets. He et al. (2005) investigated vibration analysis of multi-layered graphene sheets in which the van der Waals interaction between layers is described by an explicit formula. Behfar and Naghdabadi (2005) studied nano scale vibrational analysis of a multi-layered graphene sheet embedded in an elastic medium based on the classical plate theory. Lu et al. (2007) derived the basic equations of nonlocal Kirchhoff and Mindlin plate theories for simply supported nano-plates. Axisymmetric bending of micro/nanoscale circular plates was studied using a nonlocal plate theory by Duan and Wang (2007). Pradhan and Phadikar (2009a) presented classical and first order shear deformation plate theories for vibration of nano-plate. Their approach is based on the Navier solution and for a nano-plate with all edges simply supported. Pradhan and Phadikar (2009b) carried out vibration analysis of multilayered graphene sheets embedded in polymer matrix employing nonlocal continuum mechanics.

In-plane vibration of nano-plates was investigated by Murmu and Pradhan (2009b) employing nonlocal continuum mechanics and considering small scale effect.

Aghababaei and Reddy (2009) developed a higher order plate theory for buckling and vibration analyses of a simply supported plate accounting the small scale effect. A nonlocal plate model was developed to study the vibrational characteristics of multi-layered graphene sheets with different boundary conditions embedded in an elastic medium using finite element method (Ansari et al., 2010). Pradhan and Kumar (2010) investigated the small scale effect on the vibration analysis of orthotropic single layered graphene sheets embedded in an elastic medium. Jomehzadeh and Saidi (2011a) investigated the nonlocal three dimensional elastodynamics theory to study the vibration of nano-plates. Recently, they (2011b & 2011c) studied the nonlinear vibration of graphene sheets using classical plate theory.

In this chapter, the vibration analysis of a nano-plate is presented by considering the small scale effect. The three coupled governing equations of motion are obtained based on the nonlocal continuum theory and are decoupled into two new equations. Solving these two decoupled partial differential equations, the natural frequencies of the nano-plate with arbitrary boundary conditions are determined. Finally, a detailed study is carried out to understand the effects of boundary condition, nonlocal parameter, thickness to length and aspect ratios on the vibration characteristics of nano-plates.

Results for natural frequencies of nano-plates with arbitrary boundary conditions are given for the first time and these can serve as reference values for other numerical analysis.

2. Constitutive relations

According to nonlocal elasticity theory, the stress at a reference point X is considered to be a function of the strain field at every point X' in the body. The nonlocal stress tensor σ^{nl} at point X can be expressed as (Eringen, 1983)

$$\sigma^{nl} = \int K(|X' - X|, \tau)\sigma^{l}(X')dX' \tag{1}$$

where σ^{l} is the classical stress tensor and $K(|X' - X|)$ is the Kernel function represents the nonlocal modulus. While the constitutive equations of classical elasticity is an algebraic relation between stress and strain tensors, that of nonlocal elasticity involves spatial integrals which represent weighted averages of contributions of the strain of all points in the body to the stress at the given point. Eringen showed that it is possible to represent the integral constitutive relation in an equivalent differential form as

$$(1-\mu\nabla^2)\sigma^{nl}=\sigma^l \tag{2}$$

where $\mu=(e_0a)^2$ is nonlocal parameter, a an internal characteristic length and e_0 a constant. Also, ∇^2 is the Laplacian operator.

3. Governing equations of motion

The first order shear deformation plate theory assumes that the plane sections originally perpendicular to the longitudinal plane of the plate remain plane, but not necessarily perpendicular to the longitudinal plane. This theory accounts for shear strains in the thickness direction of the plate and is based on the displacement field

$$\begin{aligned} u &= u_0(x,y)+z\psi_x(x,y,t)\\ v &= v_0(x,y)+z\psi_y(x,y,t)\\ w &= w(x,y)\end{aligned} \tag{3}$$

where u_0 and v_0 are displacement components of the midplane, w is transverse displacement, t is time, ψ_x and ψ_y are the rotation functions of the midplane normal to x and y directions, respectively. Using the Hamilton's principle, the nonlocal bending governing equations of motion for a single layered nano-plate are obtained as follows (Pradhan and Phadikar, 2009a)

$$D(\psi_{x,xx}+\psi_{y,xy})+\frac{D(1-\nu)}{2}(\psi_{x,yy}-\psi_{y,xy})-\kappa^2Gh(\psi_x+w_{,x})=I_2(\ddot{\psi}_x-\mu\nabla^2\ddot{\psi}_x) \tag{4a}$$

$$D(\psi_{y,yy}+\psi_{x,xy})+\frac{D(1-\nu)}{2}(\psi_{y,xx}-\psi_{x,xy})-\kappa^2Gh(\psi_y+w_{,y})=I_2(\ddot{\psi}_y-\mu\nabla^2\ddot{\psi}_y) \tag{4b}$$

$$\kappa^2Gh(\psi_{x,x}+\psi_{y,y}+w_{,xx}+w_{,yy})+q(x,y,t)=I_1(\ddot{w}-\mu\nabla^2\ddot{w}) \tag{4c}$$

In above equations, dot above each parameter denotes derivative with respect to time, G is the shear modulus, $D=Eh^3/12(1-\nu^2)$ denotes the bending rigidity of the plate, E and ν Young modulus and Poisson's ratio, respectively and κ^2 the shear correction factor. Also, q is the transverse loading in z direction. Mass moments of inertia, I_1 and I_2, are defined as

$$(I_1,I_2)=\int_{-h/2}^{h/2}\rho(1,z)dz \tag{5}$$

in which ρ is the density of the plate. It can be seen that the governing equations (4) are generally a system of six-order coupled partial differential equations in terms of the transverse displacement and rotation functions.

4. Solution

In order to solve the governing equations of motion (4) for various boundary conditions, it is reasonable to find a method to decouple these equations. Let us introduce two new functions ς and φ as

$$\varsigma = \psi_{x,x} + \psi_{y,y} \tag{6a}$$

$$\varphi = \psi_{x,y} - \psi_{y,x} \tag{6b}$$

Using relations (6), the governing equations (4) can be rewritten as

$$D\varsigma_{,x} + \frac{D(1-\nu)}{2}\varphi_{,y} - \kappa^2 Gh(\psi_x + w_{,x}) = I_2(\ddot{\psi}_x - \mu\nabla^2\ddot{\psi}_x) \tag{7a}$$

$$D\varsigma_{,y} - \frac{D(1-\nu)}{2}\varphi_{,x} - \kappa^2 Gh(\psi_y + w_{,y}) = I_2(\ddot{\psi}_y - \mu\nabla^2\ddot{\psi}_y) \tag{7b}$$

$$\kappa^2 Gh(\varsigma + \nabla^2 w) + q = I_1(\ddot{w} - \mu\nabla^2\ddot{w}) \tag{7c}$$

Doing some algebraic operations on Eqs. (7), the three coupled partial differential equations (4) can be replaced by the following two uncoupled equations

$$C\nabla^2\varphi - \kappa^2 Gh\varphi = I_2(1-\mu\nabla^2)\ddot{\varphi} \tag{8a}$$

$$\begin{aligned} D\nabla^2\nabla^2 w = (1-\mu\nabla^2)\{q - \frac{D}{\kappa^2 Gh}\nabla^2 q - I_1\ddot{w} + (\frac{I_1 D}{\kappa^2 Gh} + I_2)\nabla^2\ddot{w} + I_2\} + \\ \frac{(1-\mu\nabla^2)^2}{\kappa^2 Gh}\{I_2\ddot{q} - I_1 I_2\ddddot{w}\} \end{aligned} \tag{8b}$$

where C denotes $D(1-\nu)/2$. It can be seen that the above equations are converted to the classical equations of the Mindlin plate theory when $\mu = 0$. Like the classical elasticity (Reissner, 1985), Eqs. (8a) and (8b) are called edge-zone (boundary layer) and interior equations, respectively. Also, the rotation functions ψ_x and ψ_y can be defined in terms of w and φ as

$$\begin{aligned} \kappa^2 Gh\psi_x + I_2\ddot{\psi}_x = \frac{\partial}{\partial x}[-\frac{D(1-\mu\nabla^2)}{\kappa^2 Gh}q + \frac{I_1 D(1-\mu\nabla^2)}{\kappa^2 Gh}\ddot{w} - D\nabla^2 w - \kappa^2 Ghw + \\ I_2\mu(-\frac{(1-\mu\nabla^2)}{\kappa^2 Gh}\ddot{q} - \nabla^2\ddot{w} + \frac{I_1(1-\mu\nabla^2)}{\kappa^2 Gh}\ddddot{w})] + \frac{\partial}{\partial y}[C\varphi + I_2\mu\ddot{\varphi}] \end{aligned} \tag{9a}$$

$$\begin{aligned} \kappa^2 Gh\psi_y + I_2\ddot{\psi}_y = \frac{\partial}{\partial y}[-\frac{D(1-\mu\nabla^2)}{\kappa^2 Gh}q + \frac{I_1 D(1-\mu\nabla^2)}{\kappa^2 Gh}\ddot{w} - D\nabla^2 w - \kappa^2 Ghw + \\ I_2\mu(-\frac{(1-\mu\nabla^2)}{\kappa^2 Gh}\ddot{q} - \nabla^2\ddot{w} + \frac{I_1(1-\mu\nabla^2)}{\kappa^2 Gh}\ddddot{w})] - \frac{\partial}{\partial x}[C\varphi + I_2\mu\ddot{\varphi}] \end{aligned} \tag{9a}$$

By obtaining transverse displacement and rotation functions (w, ψ_x and ψ_y), the stress components of the nano-plate can be computed by using the nonlocal constitutive relations in the following forms

$$\sigma_{xx}^{nl} - \mu\nabla^2\sigma_{xx}^{nl} = \frac{E}{1-\nu^2}(\psi_{x,x} + \nu\psi_{y,y})z \qquad \sigma_{xy}^{nl} - \mu\nabla^2\sigma_{xy}^{nl} = \frac{E}{2(1+\nu)}(\psi_{x,y} + \psi_{y,x})z$$

$$\sigma_{yy}^{nl} - \mu\nabla^2\sigma_{yy}^{nl} = \frac{E}{1-\nu^2}(\psi_{y,y} + \nu\psi_{x,x})z \qquad \sigma_{xz}^{nl} - \mu\nabla^2\sigma_{xz}^{nl} = G(\psi_x + w_{,x}) \tag{10}$$

$$\sigma_{yz}^{nl} - \mu\nabla^2\sigma_{yz}^{nl} = G(\psi_y + w_{,y})$$

Here, a rectangular plate $(a \times b)$ with two opposite simply supported edges at $x = 0$ and $x = a$ and arbitrary boundary conditions at two other edges is considered. For free harmonic vibration of the plate, the transverse loading q is put equal to zero and the transverse deflection w and boundary layer function φ are assumed as

$$w = \sum_{n=1}^{\infty} w_n(y)\sin(\beta_n x)e^{i\omega_n t} \tag{11a}$$

$$\varphi = \sum_{n=1}^{\infty} \varphi_n(y)\cos(\beta_n x)e^{i\omega_n t} \tag{11b}$$

which exactly satisfy the simply supported boundary conditions at $x = 0$ and $x = a$. In these relations, ω_n is the natural frequency of the nano-plate and β_n denotes $n\pi / a$. Substituting the proposed series solutions (11) into decoupled Eqs. (8), yields

$$\lambda_1 \frac{\partial^4 w_n(y)}{\partial y^4} + \lambda_2 \frac{\partial^2 w_n(y)}{\partial y^2} + \lambda_3 w_n(y) = 0 \tag{12a}$$

$$\lambda_4 \frac{\partial^2 \varphi_n(y)}{\partial y^2} + \lambda_5 \varphi_n(y) = 0 \tag{12b}$$

where the constant coefficients $\lambda_i\,(i = 1,..,5)$ are material constants. The above equations are two ordinary differential equations with total order of six. The solutions of Eqs. (12) can be expressed as

$$w_n(y) = C_1\sin(\eta_1 y) + C_2\cos(\eta_1 y) + C_3\sinh(\eta_2 y) + C_4\cosh(\eta_2 y) \tag{13a}$$

$$\varphi_n(y) = C_5\sinh(\eta_3 y) + C_6\cosh(\eta_3 y) \tag{13b}$$

where $C_i\,(i = 1,..,6)$ are constants of integration and parameters η_1, η_2 and η_3 are defined as

$$\eta_1 = \sqrt{\frac{\lambda_2 + \sqrt{\lambda_2^{\,2} - 4\lambda_1\lambda_3}}{2\lambda_1}} \tag{14a}$$

$$\eta_2 = \sqrt{\frac{-\lambda_2 + \sqrt{\lambda_2^2 - 4\lambda_1\lambda_3}}{2\lambda_1}} \tag{14b}$$

$$\eta_3 = \frac{\sqrt{\lambda_4\lambda_5}}{\lambda_4} \tag{14c}$$

Six independent linear equations must be written among the integration constants to solve the free vibration problem. Applying arbitrary boundary conditions along the edges of the plate at $y=0$ and $y=b$, leads to six algebraic equations. Here, three types of boundary conditions along the edges of the nano-plate in y direction are considered as

$$\text{Simply supported } (\mathbf{S}) \quad w = M_{yy} = \psi_x = 0 \tag{14a}$$

$$\text{Clamped } (\mathbf{C}) \; w = \psi_x = \psi_y = 0 \tag{14b}$$

$$\text{Free } (\mathbf{F}) \; M_{yy} = M_{xy} = Q_y = 0 \tag{14c}$$

where the resultant moments M_{yy} and M_{xy} and resultant force Q_y are expressed as

$$M_{yy} = \int_{-h/2}^{h/2} \sigma_{yy}^{nl} z\,dz \quad M_{xy} = \int_{-h/2}^{h/2} \sigma_{xy}^{nl} z\,dz \quad Q_y = \int_{-h/2}^{h/2} \sigma_{yz}^{nl}\,dz \tag{15}$$

In order to find the natural frequencies of the nano-plate, the various boundary conditions at $y=0$ and $y=b$ should be imposed. Applying these conditions and setting the determinant of the six order coefficient matrix equal to zero, the natural frequencies of the nano-plate are evaluated.

5. Numerical results and discussion

For numerical results, the following material properties are used throughout the investigation

$$E = 1.2TPa, \; \nu = 0.3, \; \kappa^2 = 5/6 \tag{16}$$

In order to verify the accuracy of the present formulations, a comparison has been carried out with the results given by Pradhan and Phadikar (2009a) for an all edges simply supported nano-plate. To this end, a four edges simply supported nano-plate is considered. The non-dimensional natural frequency parameter $\Omega = \omega a^2 \pi^4 \sqrt{I_1 / D}$ is listed in Table 1 for some nonlocal parameters. From this table, it can be found that the present results are in good agreement with the results in literature when the rotary inertia terms have been neglected. It can be also seen that the rotary inertia terms have considerable effects especially in second mode of vibration and cause the natural frequency decreases. Hereafter, the rotary inertia terms are considered in numerical results.

To study the effects of boundary condition, the nonlocal parameter (μ) and thickness to length ratio (h/a) on the vibrational behavior of the nano-plate, the first two non-dimensional frequencies are obtained for a single layered nano-plate. The results are tabulated in Tables 2-6 for five possible boundary conditions at $y=0$ and $y=b$ as clamped- clamped (**C-C**), clamped-simply (**C-S**), clamped-free (**C-F**), simply-free (**S-F**) and free-free (**F-F**).

μ	h/b		Mode 1	Mode 2
1nm	0.1	Present	0.1322 0.1332[a]	0.1994 0.2026[a]
		Pradhan (2009a)	0.1332	0.2026
	0.2	Present	0.1210 0.1236[a]	0.1673 0.1730[a]
		Pradhan (2009a)	0.1236	0.1730
2nm	0.1	Present	0.0935 0.0942[a]	0.1410 0.1432[a]
		Pradhan (2009a)	0.0942	0.1432
	0.2	Present	0.0855 0.0874[a]	0.1183 0.1224[a]
		Pradhan (2009a)	0.0874	0.1224
3nm	0.1	Present	0.0763 0.0769[a]	0.1151 0.1170[a]
		Pradhan (2009a)	0.0769	0.1170
	0.2	Present	0.0698 0.0714[a]	0.0966 0.0999[a]
		Pradhan (2009a)	0.0714	0.0999
4nm	0.1	Present	0.0661 0.0666[a]	0.0997 0.1013[a]
		Pradhan (2009a)	0.0666	0.1013
	0.2	Present	0.0605 0.0618[a]	0.0836 0.0865[a]
		Pradhan (2009a)	0.0618	0.0865

Table 1. Comparison of non-dimensional frequency parameter $\Omega = \omega a^2 \pi^4 \sqrt{I_1 / D}$ of a nano-plate with all edges simply supported ([a] Neglecting the rotary inertia terms)

Based on the results in these tables, it can be concluded that for constant h/a, the frequency parameter decreases for all modes as the nonlocal parameter μ increases. The reason is that with increasing the nonlocal parameter, the stiffness of the nano-plate decreases. i.e. small scale effect makes the nano-plate more flexible as the nonlocal model may be viewed as atoms linked by elastic springs while the local continuum model assumes the spring constant to take on an infinite value. In sum, the nonlocal plate theory should be used if one needs accurate predictions of natural frequencies of nano-plates.

μ	h/b	Mode 1	Mode 2
1nm	0.1	0.1757	0.2124
	0.2	0.1494	0.1735
2nm	0.1	0.1242	0.1502
	0.2	0.1057	0.1227
3nm	0.1	0.1014	0.1226
	0.2	0.0863	0.1002
4nm	0.1	0.0878	0.1062
	0.2	0.0747	0.0868

Table 2. First two non-dimensional frequency parameters $\Omega = \omega a^2 \pi^4 \sqrt{I_1 / D}$ of a **C-C** nano-plate

μ	h/b	Mode 1	Mode 2
1nm	0.1	0.1501	0.2049
	0.2	0.1333	0.1700
2nm	0.1	0.1062	0.1449
	0.2	0.0942	0.1202
3nm	0.1	0.0867	0.1183
	0.2	0.0769	0.0982
4nm	0.1	0.0751	0.1024
	0.2	0.0666	0.0850

Table 3. First two non-dimensional frequency parameters $\Omega = \omega a^2 \pi^4 \sqrt{I_1 / D}$ of a **C-S** nano-plate

μ	h/b	Mode 1	Mode 2
1nm	0.1	0.1273	0.1921
	0.2	0.1172	0.1615
2nm	0.1	0.0900	0.1358
	0.2	0.0829	0.1142
3nm	0.1	0.0735	0.1109
	0.2	0.0677	0.0933
4nm	0.1	0.0636	0.0960
	0.2	0.0586	0.0808

Table 4. First two non-dimensional frequency parameters $\Omega = \omega a^2 \pi^4 \sqrt{I_1 / D}$ of a **C-F** nano-plate

The influence of thickness-length ratio on the frequency parameter can also be examined by keeping the nonlocal parameter constant while varying the thickness to length ratio. It can be easily observed that as h/a increases, the frequency parameter decreases. The decrease in the frequency parameter is due to effects of the shear deformation, rotary inertia and use of term a^2/h in the definition of the non-dimensional frequency Ω. These effects are more considerable in the second mode than in the first modes.

μ	h/b	Mode 1	Mode 2
1nm	0.1	0.1136	0.1753
	0.2	0.1070	0.1531
2nm	0.1	0.0804	0.1239
	0.2	0.0756	0.1083
3nm	0.1	0.0656	0.1012
	0.2	0.0618	0.0884
4nm	0.1	0.0568	0.0876
	0.2	0.0535	0.0766

Table 5. First two non-dimensional frequency parameters $\Omega = \omega a^2 \pi^4 \sqrt{I_1 / D}$ of a **S-F** nano-plate

μ	h/b	Mode 1	Mode 2
1nm	0.1	0.1012	0.1542
	0.2	0.0964	0.1401
2nm	0.1	0.0715	0.1090
	0.2	0.0682	0.0991
3nm	0.1	0.0582	0.0890
	0.2	0.0557	0.0809
4nm	0.1	0.0506	0.0771
	0.2	0.0481	0.0701

Table 6. First two non-dimensional frequency parameters $\Omega = \omega a^2 \pi^4 \sqrt{I_1 / D}$ of a **F-F** nano-plate

To study the effect of the boundary conditions on the vibration characteristic of the nano-plate, the frequency parameters listed in a specific row of tables 1-6 may be selected from each table. It can be seen that the lowest and highest values of frequency parameters correspond to **F-F** and **C-C** edges, respectively. Thus like the classical plate, more constrains at the edges increases the stiffness of the nano-plate which results in increasing the frequency.

The effect of variation of aspect ratio (b / a) on the natural frequency of a **C-S** nano-plate is shown in Fig. 1 for various nonlocal parameters. It can be seen with increasing the aspect ratio, the natural frequency of the nano-plate decreases because of decreasing of stiffness.

In Fig. 2, the relation between natural frequency and nonlocal parameter of a square C-C nano-plate is depicted for different thickness to length ratios. It can be seen that nonlocal theories predict smaller values of natural frequencies than local theories especially for higher thickness to length ratios. Thus the local theories, in which the small length scale effect between the individual carbon atoms is neglected, overestimate the natural frequencies. The effect of boundary conditions on the natural frequency of a nano-plate is shown in Fig. 3. It can be concluded that the boundary condition has significant effect on the vibrational characteristic of the nano-plates.

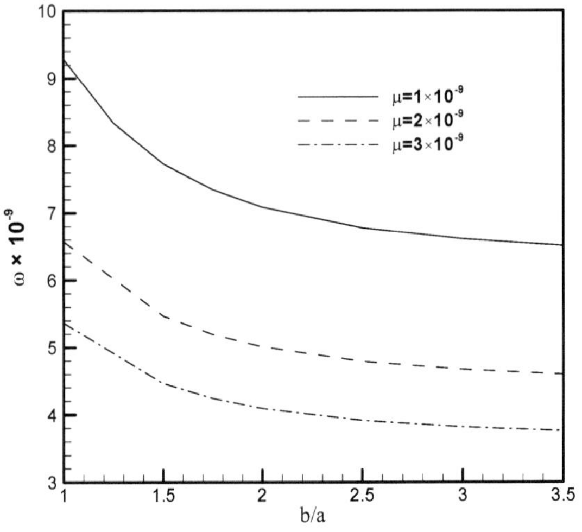

Fig. 1. Variation of natural frequency with respect to aspect ratio for a **C-S** nano-plate

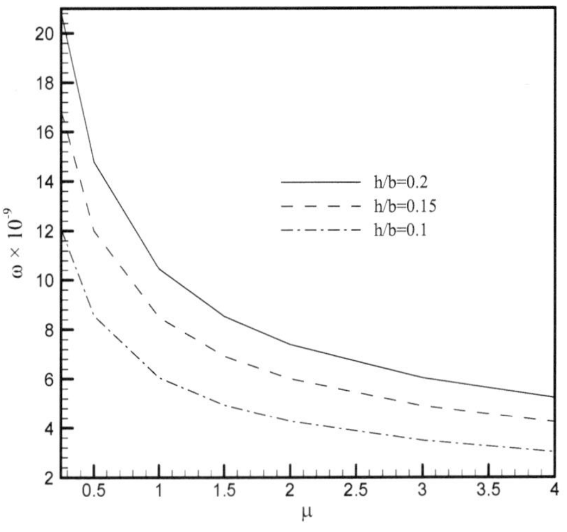

Fig. 2. Variation of natural frequency with nonlocal parameter for a **C-C** nano-plate

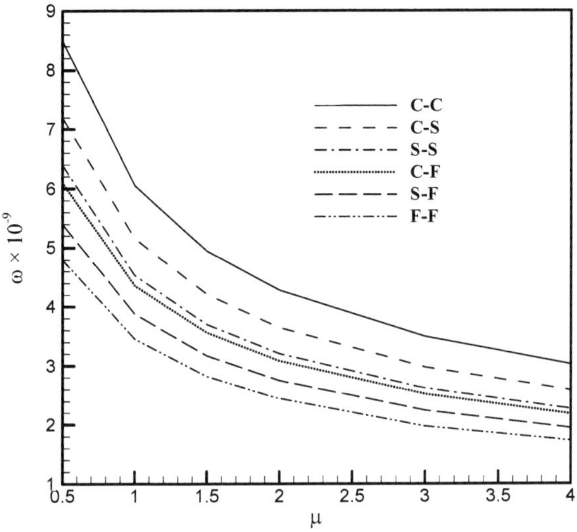

Fig. 3. Variation of natural frequency with nonlocal parameter for nano-plates with different boundary conditions at two edges

6. Conclusion

Presented herein is a variational derivation of the governing equations and boundary conditions for the free vibration of nano-plates based on Eringen's nonlocal elasticity and first order shear deformation plate theory. This nonlocal plate theory accounts for small scale effect, transverse shear deformation and rotary inertia which become significant when dealing with nano-plates. Coupled partial differential equations have been reformulated and the generalized Levy type solution has been presented for free vibration analysis of a nano-plate considering the small scale effect. The accurate natural frequencies of nano-plates have been tabulated for various nonlocal parameters, some thickness to length ratios and different boundary conditions. The effects of boundary conditions, variation of nonlocal parameter, thickness to length and aspect ratios on the frequency values of a nano-plate have been examined and discussed.

7. Acknowledgements

The authors wish to thank Iran Nanotechnology Initiative Council for its financial support.

8. References

Aghababaei R. & Reddy J.N. (2009). Nonlocal third-order shear deformation plate theory with application to bending and vibration of plates. *Journal of Sound and Vibration*, Vol. 326, pp. 277–289.

Ansari R.; Rajabiehfard R. & Arash B. (2010) Nonlocal finite element model for vibrations of embedded multi-layered graphene sheets. *Computational Materials Science*, Vol. 49, pp. 831-838.

Behfar K. & Naghdabadi R. (2005). Nanoscale vibrational analysis of a multi-layered grapheme sheet embedded in an elastic medium. *Composites Science and Technology*, Vol. 65, pp. 1159-1164.

Duan W.H. & Wang C.M. (2007). Exact solutions for axisymmetric bending of micro/nanoscale circular plates based on nonlocal plate theory. *Nanotechnology*, Vol. 18, 385704.

Eringen C. (1983). On differential equations of nonlocal elasticity and solutions of screw dislocation and surface waves. *Journal of Applied Physics*, Vol. 54, pp. 4703-4710

He X.Q.; Kitipornchai S. & Liew K.M. (2005). Resonance analysis of multi-layered grapheme sheets used as nano scale resonators. *Nanotechnology*, Vol. 16, pp. 2086-2091.

Jomehzadeh E. & Saidi A. R. (2011a) Decoupling the Nonlocal Elasticity Equations for Three Dimensional Vibration Analysis of Nanoplates. *Composite Structures*, Vol. 93, pp. 1015-1020.

Jomehzadeh E. & Saidi A. R. (2011b) A study on large amplitude vibration of multilayered graphene sheets. *Computational Materials Science*, Vol. 50, pp. 1043-1051.

Jomehzadeh E. & Saidi A. R. (2011c) Study of small scale effect on nonlinear vibration of nano-plates. *Journal of Computational and Theoretical Nanoscience*, Accepted.

Kitipornchai S.; He X.Q.; & Liew K.M. (2005). Continuum model for the vibration of multilayered graphene sheets. *Physical Review B*, Vol. 72, 075443 6 pages.

Li X.F. & Wang B.L. (2009). Vibrational modes of Timoshenko beams at small scales. *Applied Physics Letters*, Vol. 94, 1.3094130.

Lu P.; Zhang P.Q.; Lee H.P.; Wang C.M. & Reddy J.N. (2007). Non-local elastic plate theories. *Mathematical Physical and Engineering Sciences*, Vol. 463 pp. 3225-3240.

Murmu T. & Pradhan S.C. (2009a). Small-scale effect on the vibration of nonuniform nano cantilever based on nonlocal elasticity theory. *Physica E*, Vol. 41, Vol. 1451-1456.

Murmu T. & Pradhan S.C. (2009b). Small-scale effect on the free in-plane vibration of nanoplates by nonlocal continuum model. *Physica E*, Vol. 41, pp. 1628-1633.

Pradhan S.C. & Phadikar J.K. (2009a). Nonlocal elasticity theory for vibration of nanoplates. *Journal of Sound and Vibration*, Vol. 325, pp. 206-223.

Pradhan S.C. & Phadikar J.K. (2009b). Small scale effect on vibration of embedded multilayered graphene sheets based on nonlocal continuum models. *Physics Letters A*, Vol. 373, pp. 1062-1069.

Pradhan S.C. & Kumar A. (2010) Vibration analysis of orthotropic graphene sheets embedded in Pasternak elastic medium using nonlocal elasticity theory and differential quadrature method. *Computational Materials Science*, Vol. 50, pp. 239-245.

Reissner E. (1985). Reflections on the theory of elastic plates. *Applied Mechanics Review*, Vol. 38, pp. 1453-1464.

Sharma P.; Ganti S. & Bhate N. (2003). Effect of surfaces on the size-dependent elastic state of nano-inhomogeneities. *Applied Physics Letters*, Vol. 82, pp. 535-537

Wang Q. & Liew K.M. (2007). Application of nonlocal continuum mechanics to static analysis of micro- and nano-structures. *Physics Letters A*, Vol. 363, pp. 236-242.

Second Order Shear Deformation Theory (SSDT) for Free Vibration Analysis on a Functionally Graded Quadrangle Plate

A. Shahrjerdi[1] and F. Mustapha[2]
[1]*Department of Mechanical Engineering, Malayer University, Malayer,*
[2]*Department of Aerospace Engineering, Universiti Putra Malaysia, 43400 UPM, Serdang, Selangor*
[1]*Iran*
[2]*Malaysia*

1. Introduction

Studies of vibration of plates have matured and are a well-established branch of research in structural dynamics. They have a vast range of applications in engineering and technology. But not much work can be found on vibration analysis of Functionally Graded Materials (FGMs) as compared to isotropic and composite plates and shells. FGMs are those in which the volume fraction of the two or more constituent materials is varied, as a power-law distribution, continuously as a function of position along certain dimension(s) of the structure [1, 2].

From the perspective of finite element method (FEM) studies of FGM, Praveen and Reddy [3], studied the static and dynamic responses of functionally graded (FG) ceramic-metal plate accounting for the transverse shear deformation, rotary inertia and moderately large rotations in the Von-Karman sense, in which the effect of an imposed temperature field on the response of the FG plate was discussed in detail. Ng et al. [4] dealt with the parametric resonance of FG rectangular plates under harmonic in-plane loading. Ferreira and Batra [5] provided a global collocation method for natural frequencies of FG plates by a meshless method with first order shear deformation theory (FSDT). Woo et al. [6] presented an analytical solution for the nonlinear free vibration behavior of FGM plates, where the fundamental equations were obtained using the Von-Karman theory for large transverse deflection, and the solution was based in terms of mixed Fourier series. Zhao et al. [7] studied the free vibration analysis of metal and ceramic FG plates using the element-free kp-Ritz method. The FSDT was employed to account for the transverse shear strain and rotary inertia, mesh-free kernel particle functions were used to approximate the two-dimensional displacement fields and the eigen-equation was obtained by applying the Ritz procedure to the energy functional of the system. Batra and Jin [8] used the FSDT coupled with the FEM to study the free vibrations of an FG anisotropic rectangular plate with various edge conditions. Also, Batra and Aimmanee [9] studied a higher order shear and normal deformable plate theory by FEM. Many studies conducted on FGMs are related to the analysis of free vibration by applying FSDT (see [10-12] and the references there in).

Other forms of shear deformation theory, such as the third order-shear deformation theory (TSDT) that accounts for the transverse effects, have been considered. Cheng and Batra [13] applied Reddy's third order plate theory to study buckling and steady state vibrations of a simply supported FG isotropic polygonal plate [14]. Vel and Batra [14] dealt with the three-dimensional exact solution for free and forced vibrations of simply supported FGM rectangular plates using FDST and TSDT by employing the power series method. Nonlinear vibration and dynamic response of FGM plates in thermal environments were studied by Huang et al. [15] based on the higher-order shear deformation plate theory and general Von-Karman type equation. Static analysis of FG plates using TSDT and a meshless method were also presented by Ferreira et al. [16].

As for the first-order shear deformation plate theory (FSDT), the theory extends the kinematics of the classical plate theory (CPT) by relaxing the normality restriction and allowing for arbitrary but constant rotation of transverse normals. On the other hand, the second and third order shear deformation plate theory further relaxes the kinematic hypothesis by removing the straightness assumption; i.e., the straight normal to the middle plane before deformation may become cubic curves after deformation. The most significant difference between the classical and shear deformation theories is the effect of including transverse shear deformation on the predicted deflections, frequencies, and buckling loads [19].

A unified derivation of various shear-deformation models consists of Kirchhoff-Love type, Mindlin-Reissner type theory, third order theory, Layer-Wise theory and Exact-Solution. Librescu et al. [22] studied the correlation between two apparently different higher-order theories and First order transverse shear deformation theory (FSDT) of anisotropic plates. The Kirchhoff-Love assumptions were developed by Librescu and Schmidt [23]. The theory incorporates normal and shear deformation (transverse) as well as the higher-order effects, and accounts for small strains and moderate rotations of the normal.

For experimental work, shear deformation validation and compared structural theories, Stoffle [20] measured and simulated vibrations of viscoplastic plates under impulsive loading and determined how accurately the measured deformations can be calculated by the chosen constitutive and structural theories. He assumed a first-order shear deformation shell theory and applied small strains and moderate rotations and viscoplastic laws. He applied short time measurement techniques to shock tubes in order to record fast loading processes and plate deformations.

As mentioned above, shear deformation theories have been applied to consider transverse shear strains and rotation. Axisymmetric bending and stretching of functionally graded solid circular and annular plates were studied using the second-order shear deformation plate theory by Saidi and Sahraee [21]. Khdeir and Reddy [17] studied the free vibration of laminated composite plates using SSDT. Bahtuei and Eslami [18] also investigated the coupled thermoelastic response of a FG circular cylindrical shell by considering SSDT.

To the authors' knowledge, not much work has been done in the area of the dynamic stability of FG plate by using SSDT. In this study, the free vibration of FG plates (rectangular and square) by using SSDT is presented. The material properties of the plates are graded along the thickness direction according to a volume fraction power law distribution. Classical elasticity is considered and the complete governing equations are presented. Navier's method is applied to solve the equations. This work aims to investigate the effect of some basic factors such as material properties, side-to-side and side-to-thickness ratio for FG quadrangular plates on simply supported boundary conditions.

2. Gradation relations

The most commonly used models for most of the literature that express the variation of material properties in FGMs is the power law distribution of the volume fraction. According to this model, the material property gradation through the thickness of the plate is assumed to be in the following form [10]:

$$E = E(x_3) = (E_c - E_m)(x_3 / h + 1/2)^p + E_m \tag{1a}$$

$$\rho = \rho(x_3) = (\rho_c - \rho_m)(x_3 / h + 1/2)^p + \rho_m \tag{1b}$$

Here E and ρ denote the modulus of elasticity and density of FG structure, while the parameters with subscript m or c represent the material properties of a pure metal and pure ceramic plate, respectively. The thickness coordinate variable is presented by x_3 while $-\frac{h}{2} \le x_3 \le \frac{h}{2}$, where h is the total thickness of the plate as shown in Figure 1. $p \ge 0$ is the volume fraction exponent (also called grading index in this paper); $(x_3 / h + 1/2)^p$ denotes the volume fraction of the ceramic.

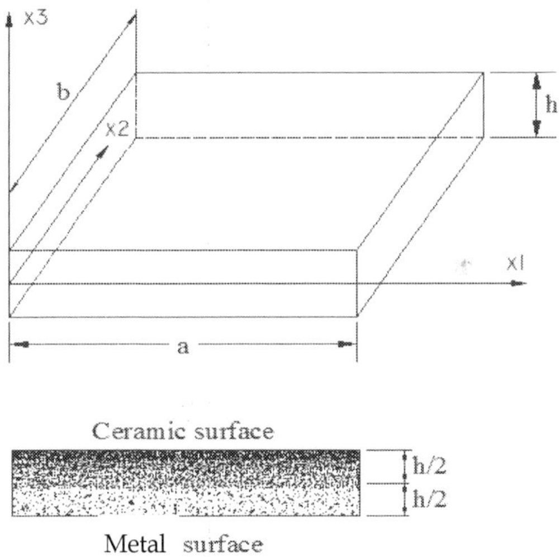

Fig. 1. Functionally graded plate.

A FG rectangular is considered as shown in Figure 1. The material in the top surface and in the bottom surface is Full-Ceramic and Full-Metal respectively, and between these two pure materials, the power law distribution of material is applied. The most well-known FGM is compositionally graded from a ceramic to a metal to incorporate such diverse properties as heat, wear and oxidation resistance of ceramics with the toughness, strength, machinability and bending capability of metals [7].

3. Elastic equations

Under consideration is a thin FG plate with constant thickness h, width, a, and length, b, as shown in Figure 1. Cartesian coordinate system (x_1, x_2, x_3) is used.

3.1 Displacement field and strains

The SSDT is based on the following representation of the displacement field:

$$u_1 = u + x_3\phi_1 + x_3^2\phi_2 \tag{2a}$$

$$u_2 = v + x_3\psi_1 + x_3^2\psi_2 \tag{2b}$$

$$u_3 = w \tag{2c}$$

Where (u_1, u_2, u_3) denote the displacement components in the (x_1, x_2, x_3) directions respectively; (u, v, w) are the displacements of a point on the mid plane $(x_1, x_2, 0)$. All displacement components $(u, v, w, \phi_1, \phi_2, \psi_1, \psi_2)$ are functions of position (x_1, x_2) and time t. The strain-displacement equations of the linear strain are given by [19].

$$\begin{Bmatrix} \varepsilon_{11} \\ \varepsilon_{22} \\ \gamma_{12} \end{Bmatrix} = \begin{Bmatrix} \varepsilon_{11}^0 \\ \varepsilon_{22}^0 \\ \varepsilon_{12}^0 \end{Bmatrix} + x_3 \begin{Bmatrix} \kappa_{11} \\ \kappa_{22} \\ \kappa_{12} \end{Bmatrix} + x_3^2 \begin{Bmatrix} \kappa_{11}^{'} \\ \kappa_{22}^{'} \\ \kappa_{12}^{'} \end{Bmatrix} \tag{3a}$$

$$\begin{Bmatrix} \gamma_{23} \\ \gamma_{13} \end{Bmatrix} = \begin{Bmatrix} \gamma_{23}^0 \\ \gamma_{13}^0 \end{Bmatrix} + x_3 \begin{Bmatrix} \gamma_{23}^1 \\ \gamma_{13}^1 \end{Bmatrix} \tag{3b}$$

where

$$\begin{aligned}
&\varepsilon_{11}^0 = \frac{\partial u}{\partial x_1}, \kappa_{11} = \frac{\partial \phi_1}{\partial x_1}, \kappa_{11}^{'} = \frac{\partial \phi_2}{\partial x_1} \\
&\varepsilon_{22}^0 = \frac{\partial v}{\partial x_2}, \kappa_{22} = \frac{\partial \psi_1}{\partial x_2}, \kappa_{22}^{'} = \frac{\partial \psi_2}{\partial x_2} \\
&\varepsilon_{12}^0 = \frac{\partial u}{\partial x_2} + \frac{\partial v}{\partial x_1}, \kappa_{12} = \frac{\partial \phi_1}{\partial x_2} + \frac{\partial \psi_1}{\partial x_1}, \kappa_{12}^{'} = \frac{\partial \phi_2}{\partial x_2} + \frac{\partial \psi_2}{\partial x_1} \\
&\gamma_{23}^0 = \psi_1 + \frac{\partial w}{\partial x_2}, \gamma_{13}^0 = \phi_1 + \frac{\partial w}{\partial x_1}, \gamma_{23}^1 = 2\psi_2, \gamma_{13}^1 = 2\phi_2
\end{aligned} \tag{4}$$

3.2 Stress-strain relations

The stress-strain relations are given by [17, 19].

$$\begin{Bmatrix} \sigma_{11} \\ \sigma_{22} \\ \sigma_{23} \\ \sigma_{13} \\ \sigma_{12} \end{Bmatrix} = \begin{bmatrix} q_{11} & q_{12} & 0 & 0 & 0 \\ q_{12} & q_{22} & 0 & 0 & 0 \\ 0 & 0 & q_{44} & 0 & 0 \\ 0 & 0 & 0 & q_{55} & 0 \\ 0 & 0 & 0 & 0 & q_{66} \end{bmatrix} \begin{Bmatrix} \varepsilon_{11} \\ \varepsilon_{22} \\ \gamma_{23} \\ \gamma_{13} \\ \gamma_{12} \end{Bmatrix} \tag{5}$$

where q_{ij} are the material constants given by

$$q_{11} = q_{22} = \frac{E}{1-\nu^2} \qquad q_{12} = \nu q_{11} \qquad q_{44} = q_{55} = q_{66} = \frac{E}{2(1+\nu)} \tag{6}$$

Hence, it follows that

$$\begin{Bmatrix} \sigma_{11} \\ \sigma_{22} \\ \sigma_{23} \\ \sigma_{13} \\ \sigma_{12} \end{Bmatrix} = \begin{bmatrix} q_{11} & q_{12} & 0 & 0 & 0 \\ q_{12} & q_{22} & 0 & 0 & 0 \\ 0 & 0 & q_{66} & 0 & 0 \\ 0 & 0 & 0 & q_{55} & 0 \\ 0 & 0 & 0 & 0 & q_{44} \end{bmatrix} \left(\begin{Bmatrix} \varepsilon_{11}^0 \\ \varepsilon_{22}^0 \\ \gamma_{23}^0 \\ \gamma_{13}^0 \\ \varepsilon_{12}^0 \end{Bmatrix} + x_3 \begin{Bmatrix} \kappa_{11} \\ \kappa_{22} \\ \gamma_{23}^1 \\ \gamma_{13}^1 \\ \kappa_{12} \end{Bmatrix} + x_3^2 \begin{Bmatrix} \kappa_{11}' \\ \kappa_{22}' \\ 0 \\ 0 \\ \kappa_{12}' \end{Bmatrix} \right) \tag{7}$$

3.3 Equations of motion

For the case of a rectangular plate, K, U and V are the kinetic, strain and potential energies of the body, respectively. The summation of the potential energy of external forces and strain energy, $U+V$, is the total potential energy, Π, of the body. Hamilton's principle for an elastic body is given by,

$$\int_{t_1}^{t_2} (\delta K - \delta\Pi)\, dt = 0 \tag{8}$$

The inertias are defined by

$$I_i = \int_{\frac{-h}{2}}^{\frac{h}{2}} \rho_0 (x_3)^i \, dx_3 \qquad (i = 0,1,2,\ldots,6) \tag{9}$$

Hamilton's principle, equation (8), along with the SSDT, given by equation (2), yields the complete form of the equilibrium equations:

$$\begin{aligned}
&\frac{\partial Q_{13}}{\partial x_1} + \frac{\partial Q_{23}}{\partial x_2} = I_0 \ddot{w} \\
&\frac{\partial M_{11}}{\partial x_1} + \frac{\partial M_{12}}{\partial x_2} - Q_{13} = I_2 \ddot{\phi}_1 + I_1 \ddot{u} + I_3 \ddot{\phi}_2 \\
&\frac{\partial L_{11}}{\partial x_1} + \frac{\partial L_{12}}{\partial x_2} - 2R_{13} = I_2 \ddot{u} + I_4 \ddot{\phi}_2 + I_3 \ddot{\phi}_1 \\
&\frac{\partial M_{22}}{\partial x_2} + \frac{\partial M_{12}}{\partial x_1} - Q_{23} = I_2 \ddot{\psi}_1 + I_1 \ddot{v} + I_3 \ddot{\psi}_2 \\
&\frac{\partial L_{22}}{\partial x_2} + \frac{\partial L_{12}}{\partial x_1} - 2R_{23} = I_2 \ddot{v} + I_4 \ddot{\psi}_2 + I_3 \ddot{\psi}_1
\end{aligned} \tag{10}$$

where N, M, L, Q and R are the stress resultants. These parameters can be represented by

$$\begin{Bmatrix} N_{11} \\ N_{22} \\ N_{12} \end{Bmatrix} = \int_{\frac{-h}{2}}^{\frac{h}{2}} \begin{Bmatrix} \sigma_{11} \\ \sigma_{22} \\ \sigma_{12} \end{Bmatrix} dx_3 \quad \begin{Bmatrix} N_{11} \\ N_{22} \\ N_{12} \end{Bmatrix} = \begin{bmatrix} A_{11}\varepsilon_{11}^0 + B_{11}\kappa_{11} + D_{11}\kappa_{11}^{'} + A_{12}\varepsilon_{22}^0 + B_{12}\kappa_{22} + D_{12}\kappa_{22}^{'} \\ A_{12}\varepsilon_{11}^0 + B_{12}\kappa_{11} + D_{12}\kappa_{11}^{'} + A_{22}\varepsilon_{22}^0 + B_{22}\kappa_{22} + D_{22}\kappa_{22}^{'} \\ A_{66}\varepsilon_{12}^0 + B_{66}\kappa_{12} + D_{66}\kappa_{12}^{'} \end{bmatrix} \tag{11a}$$

$$\begin{Bmatrix} M_{11} \\ M_{22} \\ M_{12} \end{Bmatrix} = \int_{\frac{-h}{2}}^{\frac{h}{2}} \begin{Bmatrix} \sigma_{11} \\ \sigma_{22} \\ \sigma_{12} \end{Bmatrix} x_3 dx_3 \quad \begin{Bmatrix} M_{11} \\ M_{22} \\ M_{12} \end{Bmatrix} = \begin{bmatrix} B_{11}\varepsilon_{11}^0 + D_{11}\kappa_{11} + E_{11}\kappa_{11}^{'} + B_{12}\varepsilon_{22}^0 + D_{12}\kappa_{22} + E_{12}\kappa_{22}^{'} \\ B_{12}\varepsilon_{11}^0 + D_{12}\kappa_{11} + E_{12}\kappa_{11}^{'} + B_{22}\varepsilon_{22}^0 + D_{22}\kappa_{22} + E_{22}\kappa_{22}^{'} \\ B_{66}\varepsilon_{12}^0 + D_{66}\kappa_{12} + E_{66}\kappa_{12}^{'} \end{bmatrix} \tag{11b}$$

$$\begin{Bmatrix} L_{11} \\ L_{22} \\ L_{12} \end{Bmatrix} = \int_{\frac{-h}{2}}^{\frac{h}{2}} \begin{Bmatrix} \sigma_{11} \\ \sigma_{22} \\ \sigma_{12} \end{Bmatrix} x^2{}_3 dx_3 \quad \begin{Bmatrix} L_{11} \\ L_{22} \\ L_{12} \end{Bmatrix} = \begin{bmatrix} D_{11}\varepsilon_{11}^0 + E_{11}\kappa_{11} + F_{11}\kappa_{11}^{'} + D_{12}\varepsilon_{22}^0 + E_{12}\kappa_{22} + F_{12}\kappa_{22}^{'} \\ D_{12}\varepsilon_{11}^0 + E_{12}\kappa_{11} + F_{12}\kappa_{11}^{'} + D_{22}\varepsilon_{22}^0 + E_{22}\kappa_{22} + F_{22}\kappa_{22}^{'} \\ D_{66}\varepsilon_{12}^0 + E_{66}\kappa_{12} + F_{66}\kappa_{12}^{'} \end{bmatrix} \tag{11c}$$

$$\begin{Bmatrix} Q_{13} \\ Q_{23} \end{Bmatrix} = \int_{\frac{-h}{2}}^{\frac{h}{2}} \begin{Bmatrix} \sigma_{13} \\ \sigma_{23} \end{Bmatrix} dx_3 , \begin{Bmatrix} Q_{13} \\ Q_{23} \end{Bmatrix} = \begin{bmatrix} A_{55}\gamma_{13}^0 + B_{55}\gamma_{13}^1 \\ A_{44}\gamma_{23}^0 + B_{44}\gamma_{23}^1 \end{bmatrix} \tag{11d}$$

$$\begin{Bmatrix} R_{13} \\ R_{23} \end{Bmatrix} = \int_{\frac{-h}{2}}^{\frac{h}{2}} \begin{Bmatrix} \sigma_{13} \\ \sigma_{23} \end{Bmatrix} x_3 dx_3 , \begin{Bmatrix} R_{13} \\ R_{23} \end{Bmatrix} = \begin{bmatrix} B_{55}\gamma_{13}^0 + D_{55}\gamma_{13}^1 \\ B_{44}\gamma_{23}^0 + D_{44}\gamma_{23}^1 \end{bmatrix} \tag{11e}$$

where

$$A_{ij}, B_{ij}, D_{ij}, E_{ij}, F_{ij} = \int_{\frac{-h}{2}}^{\frac{h}{2}} q_{ij}\left(1, x_3, x_3^2, x_3^3, x_3^4\right) dx_3 \tag{12a}$$

Here $A_{ij}, B_{ij}, D_{ij}, E_{ij}\, and\, F_{ij}$ are the plate stiffnesses.

$$\text{For} \begin{cases} A_{ij}, D_{ij}, F_{ij} & (i, j = 1,2,4,5,6) \\ E_{ij}, B_{ij} & (i, j = 1,2,6) \end{cases} \tag{12b}$$

By substituting equation (4) into equation (11) and then into equation (10) and also by applying definition (12), Navier's equations for FG plates are obtained as follows:

$$\begin{aligned} &A_{11}\frac{\partial^2 u}{\partial x_1^2} + A_{66}\frac{\partial^2 u}{\partial x_2^2} + (A_{12} + A_{66})\frac{\partial^2 v}{\partial x_1 \partial x_2} + B_{11}\frac{\partial^2 \phi_1}{\partial x_1^2} + B_{66}\frac{\partial^2 \phi_1}{\partial x_2^2} + D_{11}\frac{\partial^2 \phi_2}{\partial x_1^2} + \\ &D_{66}\frac{\partial^2 \phi_2}{\partial x_2^2} + (B_{12} + B_{66})\frac{\partial^2 \psi_1}{\partial x_1 \partial x_2} + (D_{12} + D_{66})\frac{\partial^2 \psi_2}{\partial x_1 \partial x_2} = I_0\ddot{u} + I_2\ddot{\phi}_2 + I_1\ddot{\phi}_1 \end{aligned} \tag{13a}$$

$$A_{12}\frac{\partial^2 u}{\partial x_1 \partial x_2}+A_{66}\frac{\partial^2 u}{\partial x_1 \partial x_2}+A_{22}\frac{\partial^2 v}{\partial x_2^2}+A_{66}\frac{\partial^2 v}{\partial x_1^2}+(B_{12}+B_{66})\frac{\partial^2 \phi_1}{\partial x_1 \partial x_2}+(D_{12}+D_{66})\frac{\partial^2 \phi_2}{\partial x_1 \partial x_2}$$
$$+B_{66}\frac{\partial^2 \psi_1}{\partial x_1^2}+B_{22}\frac{\partial^2 \psi_1}{\partial x_2^2}+D_{66}\frac{\partial^2 \psi_2}{\partial x_1^2}+D_{22}\frac{\partial^2 \psi_2}{\partial x_2^2}=I_0\ddot{v}+I_2\ddot{\psi}_2+I_1\ddot{\psi}_1 \quad (13b)$$

$$A_{55}\frac{\partial \phi_1}{\partial x_1}+A_{55}\frac{\partial^2 w}{\partial {x_1}^2}+2B_{55}\frac{\partial \phi_2}{\partial x_1}+A_{44}\frac{\partial \psi_1}{\partial x_2}+A_{44}\frac{\partial^2 w}{\partial {x_2}^2}+2B_{44}\frac{\partial \psi_2}{\partial x_2}=I_0\ddot{w} \quad (13c)$$

$$B_{11}\frac{\partial^2 u}{\partial x_1^2}+B_{66}\frac{\partial^2 u}{\partial x_2^2}+(B_{12}+B_{66})\frac{\partial^2 v}{\partial x_1 \partial x_2}+D_{11}\frac{\partial^2 \phi_1}{\partial x_1^2}+D_{66}\frac{\partial^2 \phi_1}{\partial x_2^2}$$
$$+E_{11}\frac{\partial^2 \phi_2}{\partial x_1^2}+E_{66}\frac{\partial^2 \phi_2}{\partial x_2^2}+(D_{12}+D_{66})\frac{\partial^2 \psi_1}{\partial x_1 \partial x_2}+(E_{12}+E_{66})\frac{\partial^2 \psi_2}{\partial x_1 \partial x_2} \quad (13d)$$
$$-A_{55}\left(\frac{\partial w}{\partial x_1}\right)-A_{55}\phi_1-2B_{55}\phi_2=I_2\ddot{\phi}_1+I_1\ddot{u}+I_3\ddot{\phi}_2$$

$$D_{11}\frac{\partial^2 u}{\partial x_1^2}+D_{66}\frac{\partial^2 u}{\partial x_2^2}+(D_{12}+D_{66})\frac{\partial^2 v}{\partial x_1 \partial x_2}+E_{11}\frac{\partial^2 \phi_1}{\partial x_1^2}+E_{66}\frac{\partial^2 \phi_1}{\partial x_2^2}+F_{11}\frac{\partial^2 \phi_2}{\partial x_1^2}+F_{66}\frac{\partial^2 \phi_2}{\partial x_2^2}$$
$$+(E_{12}+E_{66})\frac{\partial^2 \psi_1}{\partial x_1 \partial x_2}+(F_{12}+F_{66})\frac{\partial^2 \psi_2}{\partial x_1 \partial x_2}-2\left(B_{55}\left(\frac{\partial w}{\partial x_1}\right)+2D_{55}\phi_2+B_{55}\phi_1\right) \quad (13e)$$
$$=I_2\ddot{u}+I_4\ddot{\phi}_2+I_3\ddot{\phi}_1$$

$$(B_{12}+B_{66})\frac{\partial^2 u}{\partial x_1 \partial x_2}+B_{22}\frac{\partial^2 v}{\partial x_2^2}+B_{66}\frac{\partial^2 v}{\partial x_1^2}-A_{44}\frac{\partial w}{\partial x_2}+(D_{12}+D_{66})\frac{\partial^2 \phi_1}{\partial x_1 \partial x_2}+$$
$$(E_{12}+E_{66})\frac{\partial^2 \phi_2}{\partial x_1 \partial x_2}+D_{66}\frac{\partial^2 \psi_1}{\partial x_1^2}+D_{22}\frac{\partial^2 \psi_1}{\partial x_2^2}+E_{66}\frac{\partial^2 \psi_2}{\partial x_1^2}+E_{22}\frac{\partial^2 \psi_2}{\partial x_2^2} \quad (13f)$$
$$-A_{44}\psi_1-2B_{44}\psi_2=I_2\ddot{\psi}_1+I_1\ddot{v}+I_3\ddot{\psi}_2$$

$$(D_{12}+D_{66})\frac{\partial^2 u}{\partial x_1 \partial x_2}+D_{66}\frac{\partial^2 v}{\partial x_1^2}+D_{22}\frac{\partial^2 v}{\partial x_2^2}-2B_{44}\frac{\partial w}{\partial x_2}+(E_{12}+E_{66})\frac{\partial^2 \phi_1}{\partial x_1 \partial x_2}$$
$$+(F_{12}+F_{66})\frac{\partial^2 \phi_2}{\partial x_1 \partial x_2}+E_{66}\frac{\partial^2 \psi_1}{\partial x_1^2}+E_{22}\frac{\partial^2 \psi_1}{\partial x_2^2}+F_{66}\frac{\partial^2 \psi_2}{\partial x_1^2}+F_{22}\frac{\partial^2 \psi_2}{\partial x_2^2} \quad (13g)$$
$$-2B_{44}\psi_1-4D_{44}\psi_2=I_2\ddot{v}+I_4\ddot{\psi}_2+I_3\ddot{\psi}_1$$

It can be noted by considering zero values for ϕ_2 & ψ_2 in equations (10) and (13), the FSDT equations can be obtained [19].

4. Boundary conditions

For the case of simply supported boundary conditions of FG, as shown in Figure 2, the following relations can be written:

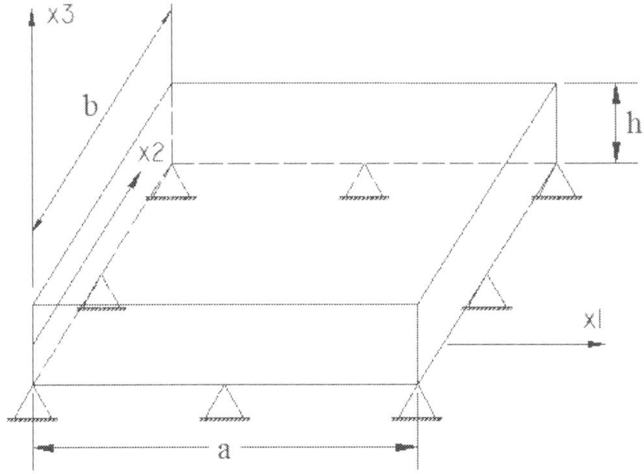

Fig. 2. Simply supported boundary condition in FG plates.

$$x_1 = 0, a \Rightarrow \begin{cases} v(0,x_2,t)=0 \\ v(a,x_2,t)=0 \\ \psi_1(0,x_2,t)=0 \\ \psi_1(a,x_2,t)=0 \\ \psi_2(0,x_2,t)=0 \\ \psi_2(a,x_2,t)=0 \end{cases}, \begin{cases} M_{11}(0,x_2,t)=0 \\ N_{11}(0,x_2,t)=0 \\ M_{11}(a,x_2,t)=0 \\ N_{11}(a,x_2,t)=0 \end{cases}, \begin{cases} w(0,x_2,t)=0 \\ w(a,x_2,t)=0 \end{cases} \tag{14a}$$

$$x_2 = 0, b \Rightarrow \begin{cases} u(x_1,0,t)=0 \\ u(x_1,b,t)=0 \\ \phi_1(x_1,0,t)=0 \\ \phi_1(x_1,b,t)=0 \\ \phi_2(x_1,0,t)=0 \\ \phi_2(x_1,b,t)=0 \end{cases}, \begin{cases} M_{22}(x_1,0,t)=0 \\ N_{22}(x_1,0,t)=0 \\ M_{22}(x_1,b,t)=0 \\ N_{22}(x_1,b,t)=0 \end{cases}, \begin{cases} w(x_1,0,t)=0 \\ w(x_1,b,t)=0 \end{cases} \tag{14b}$$

5. Method of solution

The Navier method is used for frequency analysis of a simply supported FG plate. The displacement field can be assumed to be given by:

$$u = \sum_{n=1}^{\infty} \sum_{m=1}^{\infty} u_{mn}(t) \cos \alpha x_1 \sin \beta x_2 \quad , u_{mn}(t) = Ue^{-i\omega t} \tag{15a}$$

$$v=\sum_{n=1}^{\infty}\sum_{m=1}^{\infty}v_{mn}(t)\sin\alpha x_1\cos\beta x_2 \quad ,v_{mn}(t)=Ve^{-i\omega t} \tag{15b}$$

$$w=\sum_{n=1}^{\infty}\sum_{m=1}^{\infty}w_{mn}(t)\sin\alpha x_1\sin\beta x_2 \quad ,w_{mn}(t)=We^{-i\omega t} \tag{15c}$$

$$\phi_1=\sum_{n=1}^{\infty}\sum_{m=1}^{\infty}\phi_{1mn}(t)\cos\alpha x_1\sin\beta x_2 \quad ,\phi_{1mn}(t)=\Phi_1e^{-i\omega t} \tag{15d}$$

$$\phi_2=\sum_{n=1}^{\infty}\sum_{m=1}^{\infty}\phi_{2mn}(t)\cos\alpha x_1\sin\beta x_2 \quad ,\phi_{2mn}(t)=\Phi_2e^{-i\omega t} \tag{15e}$$

$$\psi_1=\sum_{n=1}^{\infty}\sum_{m=1}^{\infty}\psi_{1mn}(t)\sin\alpha x_1\cos\beta x_2 \quad ,\psi_{1mn}(t)=\Psi_1e^{-i\omega t} \tag{15f}$$

$$\psi_2=\sum_{n=1}^{\infty}\sum_{m=1}^{\infty}\psi_{2mn}(t)\sin\alpha x_1\cos\beta x_2 \quad ,\psi_{2mn}(t)=\Psi_2e^{-i\omega t} \tag{15g}$$

where

$$\alpha=\frac{m\pi}{a} \quad ,\beta=\frac{n\pi}{b} \tag{16}$$

For natural vibrations, substituting equation (15) into the equations of motion (13), these equations reduce to the following forms:

$$[C]\begin{Bmatrix}U\\V\\W\\\Phi_1\\\Phi_2\\\psi_1\\\psi_2\end{Bmatrix}-\omega^2[M]\begin{Bmatrix}U\\V\\W\\\Phi_1\\\Phi_2\\\psi_1\\\psi_2\end{Bmatrix}=\begin{Bmatrix}0\\0\\0\\0\\0\\0\\0\end{Bmatrix} \tag{17}$$

where ω is the natural frequency and

$$\begin{array}{lll}
C_{11}=A_{11}\alpha^2+A_{66}\beta^2 & C_{21}=(A_{12}+A_{66})\alpha\beta & C_{31}=0\\
C_{12}=(A_{12}+A_{66})\alpha\beta & C_{22}=A_{66}\alpha^2+A_{22}\beta^2 & C_{32}=0\\
C_{13}=0 & C_{23}=0 & C_{33}=A_{55}\alpha^2+A_{44}\beta^2\\
C_{14}=B_{11}\alpha^2+B_{66}\beta^2 & ,C_{24}=(B_{12}+B_{66})\alpha\beta & ,\ C_{34}=A_{55}\alpha\\
C_{15}=D_{11}\alpha^2+D_{66}\beta^2 & C_{25}=(D_{12}+D_{66})\alpha\beta & C_{35}=2B_{55}\alpha\\
C_{16}=(B_{12}+B_{66})\alpha\beta & C_{26}=B_{66}\alpha^2+B_{22}\beta^2 & C_{36}=A_{44}\beta\\
C_{17}=(D_{12}+D_{66})\alpha\beta & C_{27}=D_{66}\alpha^2+D_{22}\beta^2 & C_{37}=2B_{44}\beta
\end{array}$$

$$
\begin{aligned}
&C_{41} = B_{11}\alpha^2 + B_{66}\beta^2 && C_{51} = D_{11}\alpha^2 + D_{66}\beta^2 \\
&C_{42} = (B_{12} + B_{66})\alpha\beta && C_{52} = (D_{12} + D_{66})\alpha\beta \\
&C_{43} = A_{55}\alpha && C_{53} = 2B_{55}\alpha \\
&C_{44} = D_{11}\alpha^2 + D_{66}\beta^2 + A_{55}\ , && C_{54} = E_{11}\alpha^2 + E_{66}\beta^2 + 2B_{55}\ , \\
&C_{45} = E_{11}\alpha^2 + E_{66}\beta^2 + 2B_{55} && C_{55} = F_{11}\alpha^2 + F_{66}\beta^2 + 4D_{55} \\
&C_{46} = (D_{12} + D_{66})\alpha\beta && C_{56} = (E_{12} + E_{66})\alpha\beta \\
&C_{47} = (E_{12} + E_{66})\alpha\beta && C_{57} = (F_{12} + F_{66})\alpha\beta
\end{aligned}
$$

$$
\begin{aligned}
&C_{61} = (B_{12} + B_{66})\alpha\beta && C_{71} = (D_{12} + D_{66})\alpha\beta \\
&C_{62} = B_{66}\alpha^2 + B_{22}\beta^2 && C_{72} = D_{66}\alpha^2 + D_{22}\beta^2 \\
&C_{63} = A_{44}\beta && C_{73} = 2B_{44}\beta \\
&C_{64} = (D_{12} + D_{66})\alpha\beta\ , && C_{74} = (E_{12} + E_{66})\alpha\beta \\
&C_{65} = (E_{12} + E_{66})\alpha\beta && C_{75} = (F_{12} + F_{66})\alpha\beta \\
&C_{66} = D_{66}\alpha^2 + D_{22}\beta^2 + A_{44} && C_{76} = E_{66}\alpha^2 + E_{22}\beta^2 + 2B_{44} \\
&C_{67} = E_{66}\alpha^2 + E_{22}\beta^2 + 2B_{44} && C_{77} = F_{66}\alpha^2 + F_{22}\beta^2 + 4D_{44}
\end{aligned}
\tag{18}
$$

By considering relations (18), equation (17) can be written as:

$$\left| C_{ij} - M_{ij}\omega^2 \right| = 0 \tag{19}$$

By solving equation (19) and considering appropriate values for n and m in equation (16) the fundamental frequency of a quadrangle FG plate can be obtained.

6. Validation and numerical results

6.1 Validation

The results obtained for a FG plate by applying SSDT are compared with the results obtained by using TSDT as in Ref [5] and the exact solution of [14]. The following non-dimensional fundamental frequencies in Table 1 and Table 2 are obtained by considering material properties the same as [5].

Results in Table 1 and Table 2 show that the values obtained by SSDT are greater than those from TSDT and the exact solution. This is due to the fact that the transverse shear and rotary inertia will have more of an effect on a thicker plate. For the thick plates considered in this

$h/a = 0.05$			$h/a = 0.1$			$h/a = 0.2$		
Present study	Ref. [5]	Exact [14]	Present study	Ref. [5]	Exact [14]	Present study	Ref. [5]	Exact [14]
0.0158	0.0147	0.0153	0.0621	0.0592	0.0596	0.2306	0.2188	0.2192

Table 1. Dimensionless fundamental frequency ($\bar{\omega} = \omega h\sqrt{\frac{\rho_m}{E_m}}$) of a simply supported square (Al/Zro$_2$) FG plate ($p = 1$).

case, there is insignificant difference between the result predicted by SSDT and TSDT; SSDT slightly over predicts frequencies. It can be seen that there are good agreements between our results and other results.

$p=2$			$p=3$			$p=5$		
Present study	Ref. [5]	Exact [14]	Present Study	Ref. [5]	Exact [14]	Present Study	Ref. [5]	Exact [14]
0.2292	0.2188	0.2197	0.2306	0.2202	0.2211	0.2324	0.2215	0.2225

Table 2. Dimensionless fundamental frequency ($\bar{\omega}=\omega h\sqrt{\frac{\rho_m}{E_m}}$) of a simply supported square (Al/ZrO_2) FG Plate, thickness-to-side is: $h/a=0.2$.

Material property	$E(Gpa)$	$\rho(Kg/m^3)$	ν
SUS 304, Metal	201.04	8166	0.33
Aluminum, Metal	68.9	2700	0.33
Zirconia, Ceramic	211.0	4500	0.33
Si_3N_4, Ceramic	348.43	2370	0.24

Table 3. Properties of materials used in the numerical example.

6.2 Numerical example

For numerical illustration of the free vibration of a quadrangle FG plate with Zirconia and silicon nitride as the upper-surface ceramic and aluminum and SUS 304 as the lower-surface metal are considered the same as [10]:

6.2.1 Results and discussion for the first ten modes in quadrangular FG plates

In the following Tables, free vibrations are presented in dimensionless form for square and rectangular FG plates.

Tables 4 and 5 show the dimensionless frequency in square (a=b) SUS 304/Si3N4, FG plates. It can be noted that for the same values of grading index P, the natural frequency increases with increasing mode. The effect of grading index can be shown by comparing the frequency value for the fixed value of mode and changing the values of grading index p. It can be seen that, the frequency decreases with the increase of the grading index due to the stiffness decreases from pure ceramic to pure metal.

Tables 6 and 7 show the dimensionless frequency in rectangular (b=2a) SUS 304/Si3N4, FG plates. The effect of grading index can be shown by comparing the frequency for the same value of mode and considering different values of grading index p as shown in Table 5. It is clearly visible that the frequency decreases with the increasing grading index, caused by the stiffness decreasing with increasing grading index. For the same value of p, it can be said that the natural frequency increases with increasing mode. By comparing Tables 6, 7 and 4, 5 it can be observed that for the same values of grading index and mode, the fundamental frequency in square FG plates are greater than those in rectangular FG plates and by

$m \times n$	mode	$p=0$	$p=0.5$	$p=1$	$p=2$	$p=4$	$p=6$	$p=8$	$p=10$
1x1	1	5.76	3.904	3.393	3.027	2.795	2.697	2.638	2.597
1x2	2	13.846	9.366	8.139	7.259	6.700	6.464	6.323	6.227
2x1	3	13.846	9.366	8.139	7.259	6.700	6.464	6.323	6.227
2x2	4	21.353	14.441	12.547	11.187	10.321	9.957	9.741	9.593
2x3	5	32.859	22.220	19.305	17.203	15.863	15.300	14.967	14.741
3x2	6	32.859	22.220	19.305	17.203	15.863	15.300	14.967	14.741
3x3	7	43.369	29.323	25.472	22.689	20.911	20.167	19.729	19.431
3x4	8	56.798	38.405	33.362	29.703	27.356	26.377	25.801	25.412
4x3	9	56.798	38.405	33.362	29.703	27.356	26.377	25.801	25.412
4x4	10	69.054	46.690	40.555	36.091	33.221	32.026	31.327	30.856

Table 4. Variation of the frequency parameter ($\varpi = \omega a^2 / h\sqrt{\rho_c / E_c}$) with the grading index (p) for square. $SUS\ 304 / Si_3N_4$ FG square plates ($a/h = 10, a = b$).

$m \times n$	mode	$p=0$	$p=0.5$	$p=1$	$p=2$	$p=4$	$p=6$	$p=8$	.$p=10$.
1x1	1	5.338	3.610	3.137	2.796	2.580	2.489	2.435	2.398
1x2	2	11.836	8.003	6.953	6.193	5.706	5.502	5.382	5.301
2x1	3	11.836	8.003	6.953	6.193	5.706	5.502	5.382	5.301
2x2	4	17.263	11.672	10.138	9.022	8.305	8.006	7.831	7.714
2x3	5	24.881	16.828	14.621	13.002	11.950	11.513	11.258	11.089
3x2	6	24.881	16.828	14.621	13.002	11.950	11.513	11.258	11.089
3x3	7	31.354	21.209	18.426	16.375	15.0343	14.477	14.156	13.943
3x4	8	39.180	26.508	23.041	20.471	18.770	18.062	17.656	17.388
4x3	9	39.180	26.508	23.041	20.471	18.770	18.062	17.656	17.388
4x4	10	46.020	31.141	27.067	24.036	22.020	21.181	20.702	20.387

Table 5. Variation of the frequency parameter ($\varpi = \omega a^2 / h\sqrt{\rho_c / E_c}$) with the grading index (p) for $SUS\ 304 / Si_3N_4$ FG square plates ($a/h = 5, a = b$).

$m \times n$	mode	$p=0$	$p=0.5$	$p=1$	.$p=2$.	$p=4$	$p=6$	$p=8$	$p=10$
1x1	1	3.461	2.341	2.034	1.814	1.674	1.616	1.580	1.556
1x2	2	5.338	3.610	3.137	2.796	2.580	2.489	2.435	2.39
2x1	3	10.334	6.984	6.065	5.402	4.980	4.804	4.700	
2x2	4	11.836	8.00	6.948	6.188	5.702	5.499	5.380	5.300
2x3	5	14.199	9.599	8.337	7.422	6.836	6.592	6.449	6.552
3x2	6	20.484	13.845	12.020	10.689	9.835	9.482	9.276	9.139
3x3	7	22.373	15.125	13.133	11.678	10.740	10.352	10.126	9.976
3x4	8	24.881	16.824	14.611	12.989	11.940	11.505	11.254	11.085
4x3	9	31.656	21.409	18.585	16.506	15.157	14.602	14.282	14.071
4x4	10	33.715	22.805	19.802	17.587	16.142	15.547	15.205	14.979

Table 6. Variation of the frequency parameter ($\varpi = \omega a^2 / h\sqrt{\rho_c / E_c}$) with the grading index (p) for $SUS\ 304 / Si_3N_4$ FG rectangular plate ($a/h = 5, a = 0.5 \times b$).

$m\times n$	mode	$p=0$	$p=0.5$	$p=1$	$p=2$	$p=4$	$p=6$	$p=8$	$p=10$
1x1	1	3.645	2.467	2.144	1.913	1.766	1.704	1.667	1.642
1x2	2	5.769	3.904	3.393	3.027	2.795	2.697	2.638	2.597
2x1	3	11.885	8.039	6.986	6.231	5.752	5.549	5.429	5.346
2x2	4	13.846	9.365	8.138	7.258	6.699	6.463	6.323	6.227
2x3	5	17.037	11.523	10.012	8.928	8.239	7.949	7.776	7.658
3x2	6	26.092	17.640	15.325	13.659	12.600	12.156	11.893	11.713
3x3	7	28.958	19.578	17.008	15.158	13.981	13.487	13.195	12.995
3x4	8	32.859	22.215	19.299	17.197	15.858	15.297	14.965	14.739
4x3	9	43.873	29.653	25.754	22.937	21.142	20.393	19.951	19.652
4x4	10	47.344	32.002	27.794	24.715	22.809	21.999	21.522	21.199

Table 7. Variation of the frequency parameter ($\varpi = \omega a^2 / h\sqrt{\rho_c / E_c}$) with the grading index (p) for $SUS\,304 / Si_3N_4$ FG rectangular plate ($a / h = 10, a = 0.5 \times b$).

increasing the side-to-thickness ratio, the frequency also increases. It is evident that the grading index and side-to-thickness ratio effects in frequency are more significant than the other conditions.

6.2.2 Results and discussion for the natural frequency in quadrangular FG (SUS 304/Si3N4) plates

Figures (3) and (4) illustrate the dimensionless frequency versus grading index (p), for different values of side-to-thickness ratio (a / h) and side-to-side ratio (b / a), respectively. In Figure 3, the effect of grading index (p) and side-to-thickness ratio (a/h) on dimensionless fundamental frequency of FG (SUS 304/Si3N4) plate is shown. It can be seen that the frequency decreases with increasing grading index, due to degradation of stiffness by the metallic inclusion. It can be observed that the natural frequency is maximum for full-ceramic ($p=0.0$) and this value increases with the increase of the side-to-thickness ratio, since the stiffness of thin plates is more effectively than the thick plates. It is seen that for the values (p), for $0<p<2$ the slope is greater than other parts ($p>2$). It can be said that for side-to-thickness ratios greater than twenty ($a / h > 20$), the frequencies will be similar for different values of grading index. It can be noted that the difference between frequencies in $a / h = 5$ and $a / h = 10$ are greater than differences of frequency between $a / h = 10$ and other curves for the same values of grading index p. And also it can be concluded that for $a / h > 20$, the difference between the frequencies is small for the same value of grading index.

The effect of grading index (p) and side-to-side ratio (b / a) on dimensionless fundamental frequency of FG (SUS 304/Si3N4) plate can be seen in figure 4. It can be noted that the frequency increases with the increase of the b / a since rectangular plates can be treated as a one-dimensional problem for example, beams or plate strips. It can be observed that the frequency is almost constant for different values of grading index.

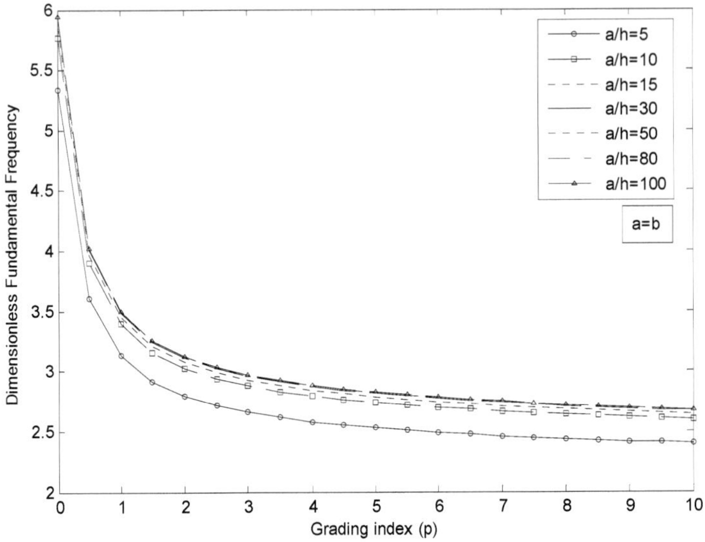

Fig. 3. Dimensionless frequency ($\varpi = \omega a^2 / h\sqrt{\rho_c / E_c}$) versus grading index ($p$) for different values of side-to-thickness ratio (a / h) in square ($b = a$) FG ($SUS\ 304 / Si_3N_4$) plates.

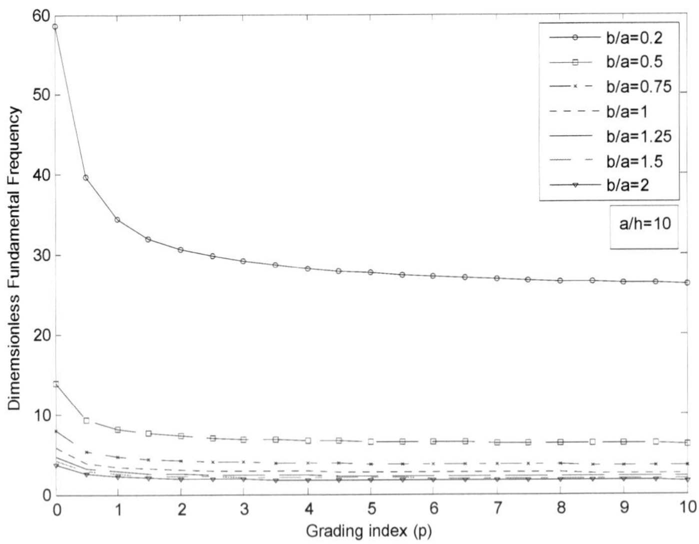

Fig. 4. Dimensionless frequency ($\varpi = \omega a^2 / h\sqrt{\rho_c / E_c}$) versus grading index ($p$) for different values of side-to-side ratio (b / a) FG ($SUS\ 304 / Si_3N_4$) plates when $a / h = 10.0$

Figures (5) and (6) show variation of dimensionless fundamental frequency of FG (SUS 304/Si3N4) plate with side-to-thickness ratio (a/h), for different values of grading index (p) and side-to side ratio (b/a), respectively.

It is seen from figure 5, the fundamental frequency increases with the increase of the value of side-to-thickness ratio (a/h). It is shown that the frequency decreases with the increase of the values of side-to-side (b/a). It can be noted that the slope of frequency versus side-to-thickness ratio (a/h) for part $5 < a/h < 10$ is greater than those in another part ($a/h > 10$).

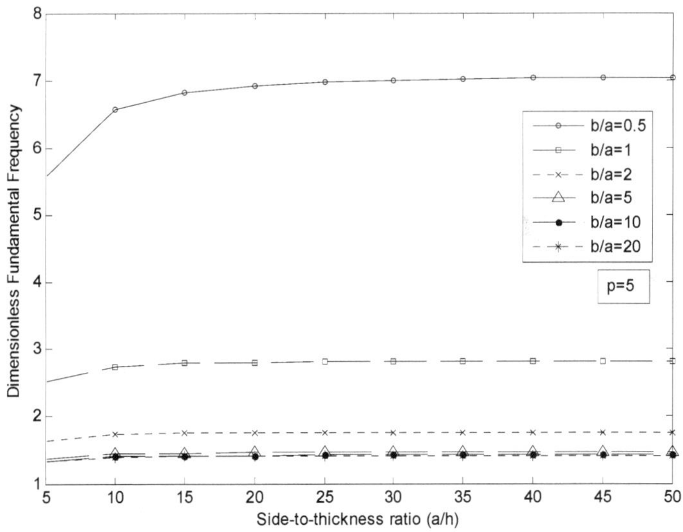

Fig. 5. Dimensionless frequency ($\varpi = \omega a^2 / h\sqrt{\rho_c / E_c}$) versus side-to-thickness ratio (a/h)for different values of side-to-side ratio (b/a) FG ($SUS\ 304 / Si_3N_4$) plates when $p = 5$.

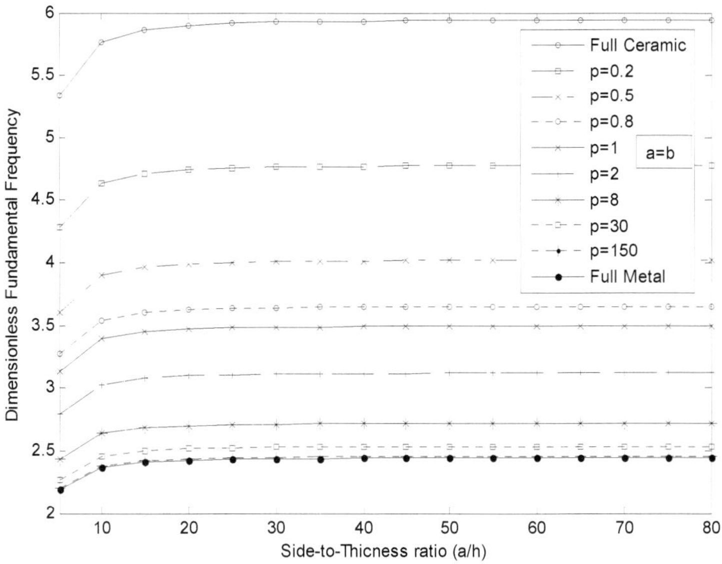

Fig. 6. Dimensionless frequency ($\varpi = \omega a^2 / h\sqrt{\rho_c / E_c}$) versus side-to-thickness ratio (a / h) for different values of grading index (p) in square ($b = a$) FG ($SUS\ 304 / Si_3N_4$) plates.

The variation of frequency with side-to-thickness ratio (a / h) for different values of grading index (p) is presented in Figure 6. As expected, by increasing the value of grading index (p) the values of frequency decrease due to the decrease in stiffness. Similarly, in figure (5) while the $5 < a / h < 10$, the slope is greater than another ratios. It can be noted that for the values of grading index $p > 30$, the results for frequency are similar.

Figures 7 and 8 present the variation of dimensionless frequency of FG (SUS 304/Si3N4) plate versus side-to-side ratio (b / a) for different values of grading index (p) and side-to-thickness ratio (a / h), respectively.

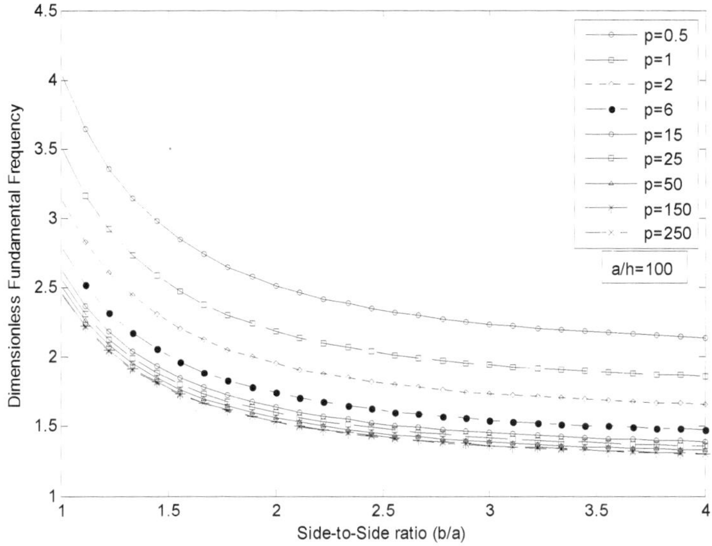

Fig. 7. Dimensionless frequency ($\varpi = \omega a^2 / h\sqrt{\rho_c / E_c}$) versus side-to-side ratio ($b = a$) for different values of grading index (p) FG ($SUS\ 304 / Si_3N_4$) plates when $a / h = 100$.

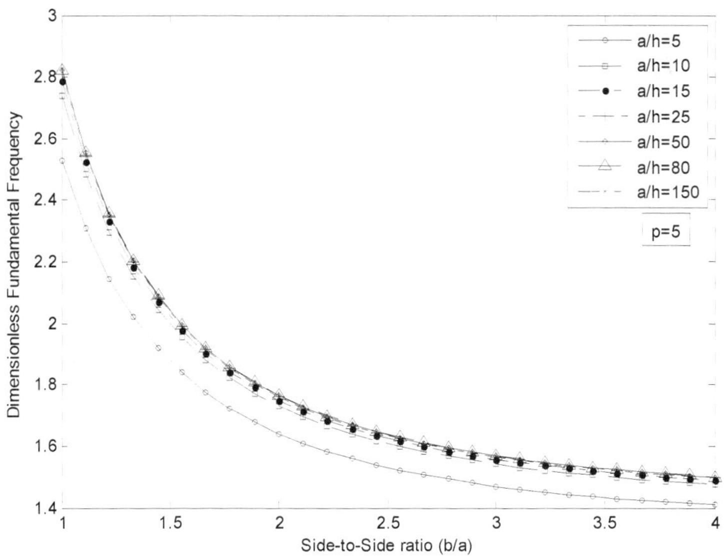

Fig. 8. Dimensionless frequency ($\varpi = \omega a^2 / h\sqrt{\rho_c / E_c}$) versus side-to-side ratio (b / a) for different values of side-to-thickness ratio (a / h) FG ($SUS\ 304 / Si_3N_4$) plates when $p = 5$.

In figure 7, it is shown that the frequency decreases with the increase of the value of side-to-side ratio (b/a) for all values of grading index (p). It is seen that the frequencies for FG quadrangular plates are between that of a full-ceramic plate and full-metal plate. As expected the frequencies in a full-ceramic plate are greater than those in a full-metal plate.
The results for dimensionless frequency versus side-to-side ratio (b/a) for different values of side-to-thickness ratio (a/h) in FG plate while grading index $p=5$ are shown in figure 8. It is seen that by increasing the value of b/a, the frequency decreases for all values of a/h. It can be noted for $a/h>10$ the results are similar.

7. Conclusions

In this chapter, free vibration of FG quadrangular plates were investigated thoroughly by adopting Second order Shear Deformation Theory (SSDT). It was assumed that the elastic properties of a FG quadrangular plate varied along its thickness according to a power law distribution. Zirconia and Si3N4 were considered as a ceramic in the upper surface while aluminum and SUS304 were considered as metals for the lower surface. The complete equations of motion were presented using Hamilton's principle. The equations were solved by using Navier's Method for simply supported FG plates.
Some general observations of this study can be deduced here:

- The decreasing slope of the fundamental frequency for $0<p<2$, is greater than another part ($p>2$) for all values of side-to-thickness ratio (a/h) in square FG plate.
- It was found that the fundamental frequency of the FG plate increases with the increase of the value of side-to-side ratio (b/a).
- For FG plates, the slope of increasing frequency versus side-to-thickness (a/h) when $5<a/h<10$ is greater than another part $(a/h>10)$ for any value of grading index and side-to-side ratio.
- The fundamental frequency versus side-to-side ratio (b/a) for FG quadrangular plates are between those of a full-ceramic plate and full-metal plate when $a/h=10$.

From the numerical results presented here, it can be proposed that the gradations of the constitutive components are the significant parameter in the frequency of quadrangular FG plates.

8. Acknowledgement

The authors would like to thank Universiti Putra Malaysia for providing the research grant (FRGS 07-10-07-398SFR 5523398) for this research work.

9. References

[1] Reddy JN. Analysis of functionally graded plates. Int. J. Numer Meth Eng 2000;47:663-684.
[2] Suresh S, Mortensen A. Fundamentals of functionally graded materials. London: IOM Communications Limited, 1998.
[3] Praveen GN, Reddy JN. Nonlinear transient thermoelastic analysis of functionally graded ceramic-metal plates. Int. J. Solids Struct 1998;35(33):4457-4476.

[4] Ng TY, Lam KY, Liew KM. Effects of FGM materials on the parametric resonance of plate structure. Comput. Meth Appl. Mech. Eng 2000;190:953-962.

[5] Ferreira AJM, Batra RC, Roque CMC, Qian LF, Jorge RMN. Natural frequencies of functionally graded plates by a meshless method. Comp Struct 2006;75:593–600.

[6] Woo J, Meguid SA, Ong LS. Nonlinear free vibration behavior of functionally graded plates. J. Sound Vibr 2006;289:595–611.

[7] Zhao X, Lee YY, Liew KM. Free vibration analysis of functionally graded plates using the element-free kp-Ritz method. J. Sound Vibr 2008.

[8] Batra RC, Jin J. Natural frequencies of a functionally graded anisotropic rectangular plate. J. Sound Vibr 2005;282:509–516.

[9] Batra RC, Aimmanee S. Vibrations of thick isotropic plates with higher order shear and normal deformable plate theories. Comput Struct 2005;83:934–955.

[10] Bayat M, Saleem M, Sahari BB, Hamouda AMS, Mahdi E. Thermo elastic analysis of a functionally graded rotating disk with small and large deflections. Thin-Walled Struct 2007;45:677–691.

[11] Bayat M, Sahari BB, Saleem M, Ali A, Wong SV. Thermo elastic solution of a functionally graded variable thickness rotating disk with bending based on the first-order shear deformation theory. Thin-Walled Struct 2008.

[12] Heidary F, M. Reza Eslami MR. Piezo-control of forced vibrations of a thermoelastic composite plate. Comp Struct 2006;74(1):99-105

[13] Cheng ZQ, Batra RC. 2000, exact correspondence between eigenvalue of membranes and functionally graded simply supported polygonal plates. J. Sound Vibr 2000;229(4):879-895.

[14] Vel SS, Batra RC. Three-dimensional exact solution for the vibration of functionally graded rectangular plates. J. Sound Vibr 2004;272: 703-30.

[15] Huang XL, Shen H. Nonlinear vibration and dynamic response of functionally graded plates in thermal environments. Int J. Solids Struct 2004;41:2403–2427.

[16] Ferreira AJM., Batra RC, Roque CMC, Qian LF, Martins PALS. Static analysis of functionally graded plates using third-order shear deformation theory and meshless method. Comp Struct 2005;69:449–457.

[17] Khdeir AA, Reddy JN. Free vibrations of laminated composite plates using second-order shear deformation theory. Comp Struct 1999;71:617-626.

[18] Bahtui A, Eslami MR. Coupled thermoelasticity of functionally graded cylindrical shells. Mech Res Commun 2007; 34(1):1-18.

[19] Reddy JN. Theory and Analysis of Elastic Plates and Shells. New York: CRC Press; 2007.

[20] Stoffel, M. (2005). Experimental validation of simulated plate deformations caused by shock waves, Math. Mech., 85(9):643 - 659.

[21] Saidi Ali Reza, Sahraee Shahab, (2006). Axisymmetric solutions of functionally graded circular and annular plates using second-order shear deformation plate theory, ESDA2006-95699, 8th Biennial ASME Conference on Engineering Systems Design and Analysis, Torino, Italy.

[22] Librescu,L., Khdeir, A.A., Reddy, J.N. (1987). Comprehensive Analysis of the State of Stressof Elastic Anisotropie Flat Plates Using Refined Theories, *Acta* Mechanica, 70:57-81.

[23] Librescu, L., Schmidt, R. (1988).Refined Theories of Elastic Anisotropic Shell Accounting for Small Strain and Moderate Rotations, Int. J. Non-Linear Mechanics, 23(3):217-229.

Part 2

Vibrations Analysis for Machine Maintenance

5

Maintenance of Reducers with an Unbalanced Load Through Vibration and Oil Analysis Predictive Techniques

Aparecido Carlos Gonçalves, Daniel Fabiano Lago and Maria da Consolação Fonseca de Albuquerque
University of State of Sao Paulo, College of Engineering of Ilha Solteira, Department of Mechanical Engineering
Brazil

1. Introduction

Among the techniques for Predicitive Maintenance of Reducers, the most used are the oil analysis and vibration analysis, and the simultaneous use of both of these provides accurate results about the condition of a system under test.

The principle of the vibration analysis is based on the idea that the structures of the machines excited by the dynamic efforts (power action) give vibrational signs, whose frequency is equal to the frequency of the exciting agents. An imbalance in a machine component will cause increased vibration, once it causes an imbalance in the system and consequent increase in the power. Thus, observing the progression of the level of vibration, it is possible to obtain information on the state of the machine (Gonçalves and Campos-Silva, 2011).

The oil analysis enables identifying the first signs of wear of a component. The identification begins with studying the amount of particles, size, shape and composition, which provide accurate information on the conditions of the moving surfaces without having to disassemble the entire set that these parties belong to. These solid particles are generated by the dynamic friction between the parts in contact. According to the study of these particles the wear situations of the set can be related and attributed to physical and chemical conditions, (Barraclough et al, 1999), (Anderson et al, 1999). The oil analysis is achieved through laboratory techniques involving, reagents, instruments and equipment..

2. Analysis of the lubricants

The lubrication is introduced between two sliding solids by the addition of a lubricant at the interface, in order to reduce friction and wear, remove heat and particles generated by the contact.

Normally, the lubricants are liquids, but can also be solid, and pasty and gaseous, according to its physical state. Liquid lubricants are characterized by their liquid viscosity, but other properties are also important. Lubricating oils have names designating these properties. This type of lubricant can be subdivided into: mineral oils, fatty oils, composite oil, aditivated oils and synthetic oils..

2.1 Viscosity

Viscosity is the most important property of lubricating oils and essentially it can be defined as the flow resistance that fluids present. It is defined as the shear stress in a plane of the fluid per unit of normal velocity gradient to the plane. Viscosity can be expressed in terms of kinematical viscosity (mm^2/s or cSt) or absolute viscosity (dynamics) whose unit is the Pa.s. Hutchings (1992) defines the Newtonian viscosity of fluids, in terms of shear deformation by:

$$\tau = \eta \cdot \frac{\partial \gamma}{\partial t} \tag{1}$$

τ = shear stress [Pa],
η= dynamic viscosity [Pa.s],
$\frac{\partial \gamma}{\partial t}$ = shear deformation rate
The kinematical viscosity Z is defined according to Equation 2.

$$Z = \frac{\eta}{\rho} \tag{2}$$

Where:
Z = viscosity in cSt or mm^2/s;
ρ = specific mass.

2.2 Viscosity monitoring

Monitoring viscosity is an important component of many programs for the analysis of oil. Even small changes in viscosity can cause major damage to the lubrication. The Limits of typical industrial oils are set at 5% for precaution, and 10% for critical situations, although applications at high loads and extremely critical systems should also have an alarm system (POA, 2002).

Significant reduction of viscosity	Significant increasing of viscosity
• Loss of the oil film causing excessive wear; • Increase of mechanical friction causing excessive consumption of energy; • Generation of heat due to mechanical friction; • Internal or external leaks; • Increased sensitivity for contamination of particle due to the reduction of the oil film; • Failure of oil film for high temperatures, high loads or during starting and stopping.	• Excessive generation of heat resulting in the oxidation of oil, sludge and development of varnish; • Gaseous cavitation due to inadequate flow of oil to pumps and bearings; • Lack of lubrication due to inadequate oil flow; • Oil whipping in the radial bearing; • Excessive consumption of energy to overcome the friction of fluid; • Poor demulsibility; • Weak pumping during cold starting.

Table 1. Depicts the effects of using a lubricant with inappropriate viscosity.

When a significant change in viscosity is observed, the root cause of the problem should always be investigated and corrected. Changes in the viscosity can result from a change in the basic chemistry of the oil (a change in the molecular structure of the oil), or due to ingression of contaminants. Change in viscosity requires additional testing, such as: number of acids (AN) and infrared spectroscopy with the Fourier transform (FTIR) to confirm the incipient oxidation; contaminant tests to identify the presence of water or soot, or another less commonly used assay, which is the gas chromatography test (GC) to identify any changes in the basic chemistry of the oil.

2.3 Wear in lubricated system

Wear can be defined as a progressive loss of material, as a result of mechanical interaction between two surfaces in contact, lubricated or not. In general these areas are in relative movement (sliding or slipping) and with applied loads.

Various authors characterize wear mechanisms differently but, according to the Modern Tribology Handbook (2001) there are 4 main forms of wear: adhesive, abrasive, corrosive and by fatigue, as well as some sideline cases often classified as wear forms. Oxidation, erosion, erosion by cavitation and impact, are sometimes classified as types of wear, although Rabinowicz (1995) finds that in fact none of them are forms of wear.

These four types of damage are shown in Figure 1.

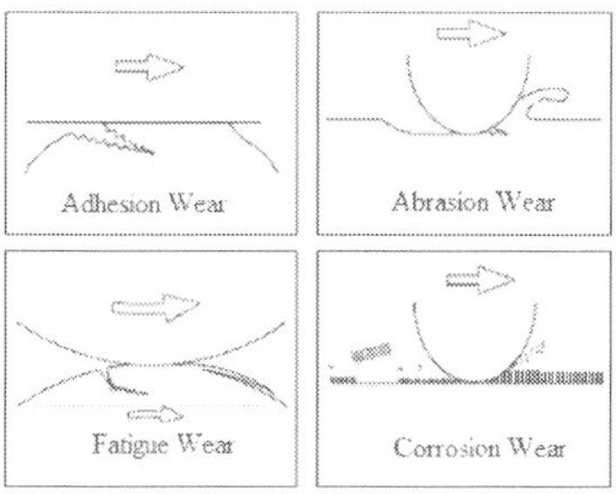

Fig. 1. Schematic Figures of four types of wear (Kato and Adashi, 2001).

These types of wear will provide precise information during the lubricant's analysis, as the particles generated mix with this analyzed lubricant.

2.4 Physical tests

The most common physical tests used with spectrographic programs and analysis programs of wear metals are: viscosity, total acid number (TAN) and determination of the water rate.

The method ASTM D445 is used for identification of viscosity, the method ASTM D 974 or D 664 is to determine the total number of acids and ASTM D 1744 to determine the water concentration by titration(Gonçalves, Padovese, 2010).

In cases where the water is at levels above 0.05 vol.%, infrared spectrography can be used. Although the limits of control for each of these parameters need to be adjusted depending on the type of lubricant and equipment, variations in viscosity of ± 10%, TAN greater than 3 mg/g, and water exceeding 100 to 500 ppm are usually sufficient for an intervention or at least for further investigation.

2.5 Spectrographic analysis of metals

The spectrographic methods include atomic absorption (AA), atomic emission spectrography (AES), inductive coupled plasma emission spectroscopy (ICPE), and X ray fluorescence (XRF). Of these methods, AES and ICPE, which are based on the detection of light emitted by the elements, are the most popular because of cost, speed, and other factors (Kimura and Gonçalves, 2009).

The spectrographic analysis of metals determines the concentration of metals and particles of up to 10 microns in size, such as moderate wear (benign sliding) and the advanced stages of fatigue, since in these wear modalities the predominant distribution of particles is within the detectable scale (<10 microns).

2.6 Particle count

The particle count is the monitoring of the number of particles of a given size by fluid volume, it is used as a preliminary monitoring tool combined with other analytical methods. The counting of particles and direct reading ferrography of direct-reading detect the onset of severe wear with a rapid increase in the quantity and size of the particles. The counting of particles detects all the particles, given that the direct reading ferrography indicates only particles of ferrous wear.

Many sensitive optical instruments are used in the counting of the number of particles in different size ranges. This counting informs the number of particles larger than a given size found in a specified volume of fluid.

For the counting of particles the standard SAE AS 4059, NAS 1638 (National Aerospace Agency), or ISO can be used (Gonçalves and Padovese, 2010).

The ISO 4406 norm (International Standard Organization) is the most widely used. Most of the versions commonly used of this technical standard refer to the number of particles larger than 4, 6, and 14 micrometers in 1 ml of fluid. The number of particles 4+ and 6+ are used as a particles point of reference. The 14+ size indicates the amount of large particles present, which contribute greatly to the possible catastrophic failure of the machine.

Figure 2 and Table 2 below represent a measurement example, where the result is obtained by an ISO code. Table 2.6 shows the various numbers of ISO 4406 code.

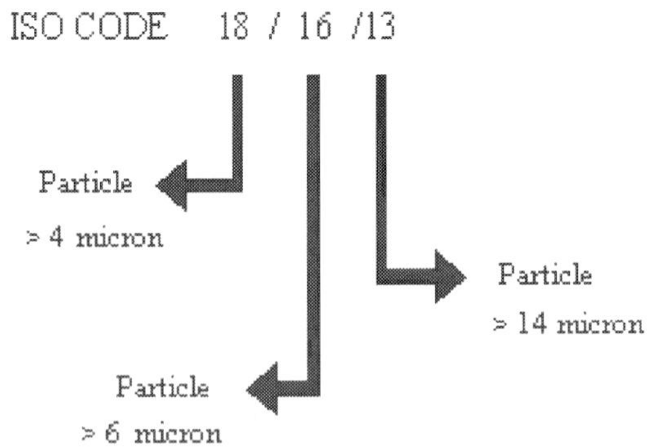

Fig. 2. Example of an obtained ISO Code, (Parker Catalog, 2000).

Range Number	Micron	Actual particle Count range (por ml)
18	4+	1,300 - 2,500
16	6+	320 - 640
13	14+	40 - 80

Table 2. Classification and definition by ISO (Parker Catalog,2000).

2.7 Ferrography

The ferrographic techniques are divided into two levels of analysis. A quantitative one, which is an evaluation technique of the wear conditions of the components of a machine, by quantifying the particles in suspension in the lubricant, and an analytical one, which uses the observation of particles in suspension in the lubricant.

The analytical ferrography is also known as direct reading ferrography. This one measures the concentration of ferrous particles in a fluid sample. With this technique one can get information on the degree of severity of wear in the machine under analysis.

To establish exact guidelines for the oil condition, samples are regularly taken from carefully selected positions of the machine system, preferably during normal operation (Lockwood and Dalley, 1992). In the ferrographic examination of direct reading the optical density is used to quantitatively measure the concentration of the wear particles in lubricating oil or in a hydraulic fluid. The particles are classified according to their sizes in DL (large particles, larger than 5 mm) (large particles), and DS (small particles, smaller than 5mm) (small particles). The wear particle concentration (WPC) and the percentage of large particles (PLP) are derived like this.

The following are equations to calculate the WPC and PBP:

$$WPC = DL+DS \quad (3)$$

$$PLP = [(DL - DS) / (DL + DS)] * 100 \quad (4)$$

The RPD (rotary particle depositor) is a type of device for direct ferrography where an index that represents the magnetic density in a given volume of lubricant is measured.

When the quantitative ferrography indicates an abnormal wear tendency, the analytical ferrographic techniques can be used to specifically identify the nature of the machine's potential problems. This enables a thorough study of particles whose size is between 1 and 250 μm (Arato, 2004). A ferrogram is built and then it can be analyzed aided by optical and electron microscopy, identifying the morphology of the particles, hence identifying any anomaly.

The ferrograms consist of transparent slides where the magnetic particles are deposited, they are separated by sizes by means of a magnetic field, other particles are randomly placed in the "barriers" formed by magnetic particles.

A first type of ferrogram is obtained by passing a flow of diluted lubricant on a plaque by gravity (the plaque is inclined). The plaque is placed on a magnet that attracts the ferrous particles and allows its adherence to the plaque, as illustrated in Figure 3.

Due to the magnetic field the particles are lined up in horizontal chains along the plaque, the larger particles are deposited first and there is a size decrease along the plaque. The non-ferrous particles are randomly placed throughout the plaque. The absence of ferrous particles actually reduces the efficiency of the analysis of non-ferrous particles.

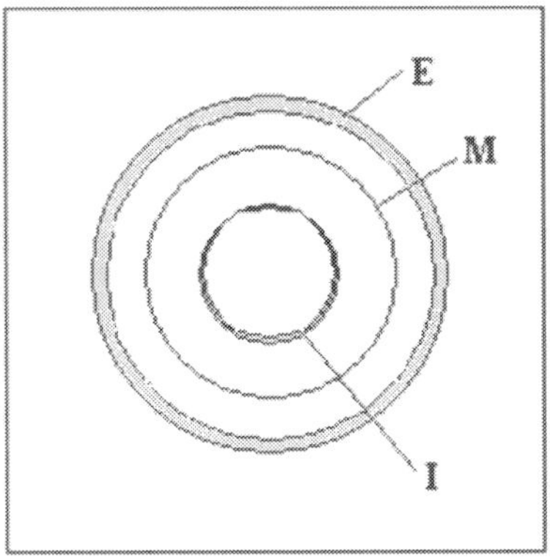

Fig. 3. Obtaining the ferrogram.

3. Analysis of vibrations

Vibration analysis is based on the idea that machine structures, excited by dynamics efforts, give vibratory signs whose frequency is identical to those of the efforts that provoked them; and the global measure in some point is the sum of vibratory answers of the structure to the different excitator efforts (Wang and Willing, 1995).

It is possible, with captors placed in private points, to register the transmitted vibration to the machine by the structures, and with its analysis, to identify the origin of the efforts which it is submitted. Vibration monitoring provides information about the macroscopic behavior inside the machine (Diana and Chelif, 2005).

In that way, as soon as it is obtained, the "vibration signature" of the machine when it is new or reputed as in good state, it will be possible for comparison , to appreciate the evolution of its state and to identify the appearance of new dynamic efforts, serial to a degradation in development process (Azoutsev, 1998).

The measure of a vibration transmitted by a structure under the effect of dynamic efforts will be function of multiple parameters:

a. Mass, rigidity and damper coefficient of the structure that transmits the vibrations;
b. Characteristics of machine fixation on the ground that opposes reactions to vibrations and modifies the intensity;
c. Position of measure point;
d. Position and fixation of the sensor (captor) on the machine;
e. Characteristic of sensor;
f. Pre amplification and transmission of the sign;
g. Performance of analyzed apparels;
h. Rotation and absorbed potency;
i. State of the connection of the cinematic chain (alignment, balancement, engagements, rolling, etc);

In a spectrum all the components of a vibratory level are represented under the form of "picks" and it can follow individually a variation of amplitude, without the masking effect that occurs in a global measure. Figure 4 shows an example of a spectrum.

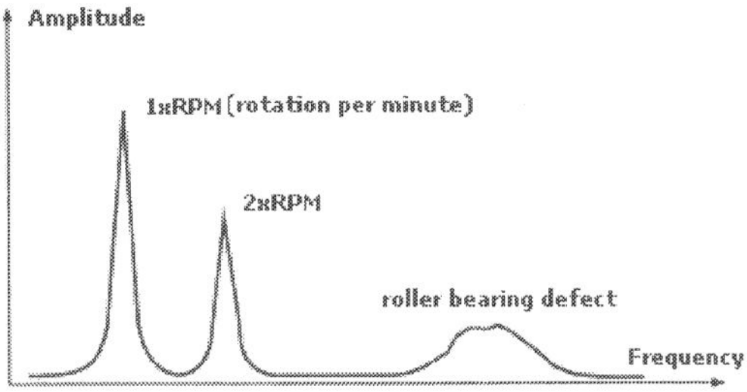

Fig. 4. Typical example of a roller bearing with problem (Mirshawaka, 1991).

The measure of the global value is an approximate analysis method of the sign that makes abstractions of the frequential parameters, expressing the evaluated amplitude in different ways. That takes into account:

a. The pick to pick value. It is measured the maximum amplitude of the fundamental wave that is useful, for example, when the vibratory displacement of a machine is critical in relation to restrictions of maximum load or clearance;
b. Crest value (or pick value). It is an important measure to indicate, for example, the level of a shock of short duration; and
c. Effective value. Measurement that takes into account the evaluation of the value of harmonic components directly related to the vibration energy content.

From the analysis of vibration signals it is possible to make decisions whether or not to intervene in the operation of the machine, so that it can be "available" as much as possible, reducing maintenance costs, time-stop of the machine, reducing the stock and improving safety, among others.

Thus, it is possible from the vibration signs at certain points of the equipment, to identify the emergence of new dynamic efforts or the abrupt increase of the magnitude of the response, which are indicators of the appearance of defects or deterioration of the operation.

Vibration data can be measured with accelerometers and processed by a signal analyzer. The signal can be analyzed in the time domain or in the frequency.

3.1 Description of time domain

A vibration signal can be presented by constructing a graph for magnitude values of the signal as a function of time, from a given moment regarded as zero time. The vibration magnitude can be represented by acceleration, velocity or movement.

The wave forms are analyzed by comparison with the wave forms previously collected, and observed by repetitive impulses that can report the frequencies of the bearings, the gears or other components. In Figure 5, the vibration signal caused by imbalance is the dominant signal. It has high magnitude when compared with the defects of bearing or gears. For this reason, the lower magnitude of the wave form is superimposed on the waves caused by imbalance (Green, 2003).

The vibration elements can be divided depending on the period of repetition, finite or infinitely long, in: periodic vibrations, random vibrations and transient vibrations.

- Periodic Vibrations - Vibrations that are repeated according to a given period of time.
- Random Vibrations - Vibrations that are unpredictable as to its instant value, for any future moment.

- Transitional Vibrations - Vibrations that exist only in a limited space in time, and null at any other time.

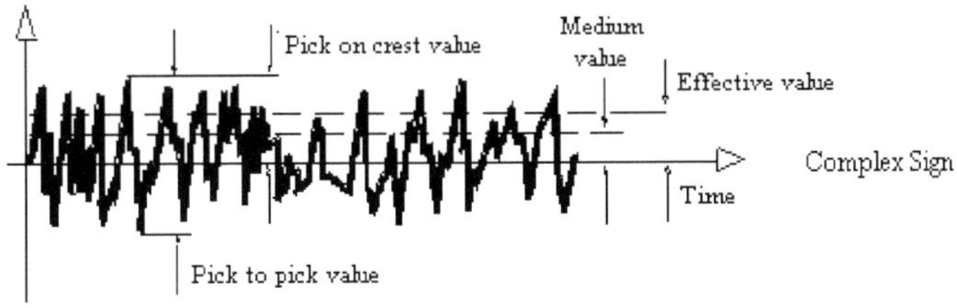

Fig. 5. Vibration Signal in time domain (Arato,2004).

3.2 Description of the frequency domain

The fast Fourier transform (FFT) can derive a wave form in time and present it in the frequency domain as shown in Figure 6. This process is the breaking of all vibrational signals into individual components of the vibration signal and plotting it in a frequency scale. This signal in the frequency domain is called frequency spectrum and provides valuable information about the condition of a machine.

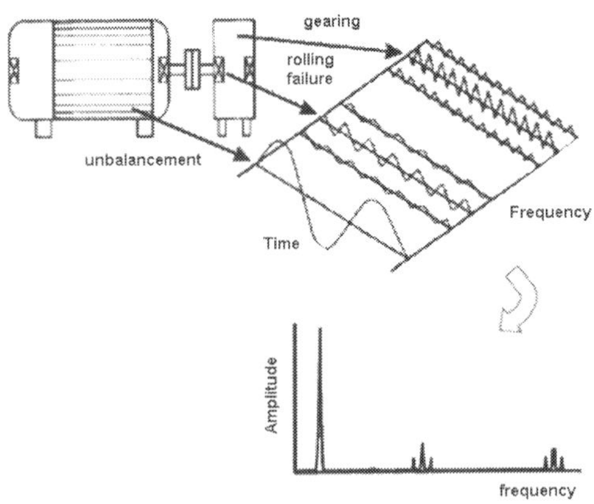

Fig. 6. Vibratory Signal in the frequency domain(Green, 2003)

The frequency spectra are used to obtain information that help determine the location of the problem, the cause of the problem and the time for the problem to become critical. This depends on the type of machine and is always relative to the level of vibration of the machine's proper operation. The frequency at which the vibration occurs indicates the type of failure and it provides an indication of what is causing the failure.

In FFT transformation, a small section is extracted from the time signal (the so-called time window) and the frequency spectrum is calculated using the FFT algorithm. During this process the instrument assumes that the signal in this time window (time data set) is continuously periodical, that is, it is repeated over time, as seen in Figure 7.

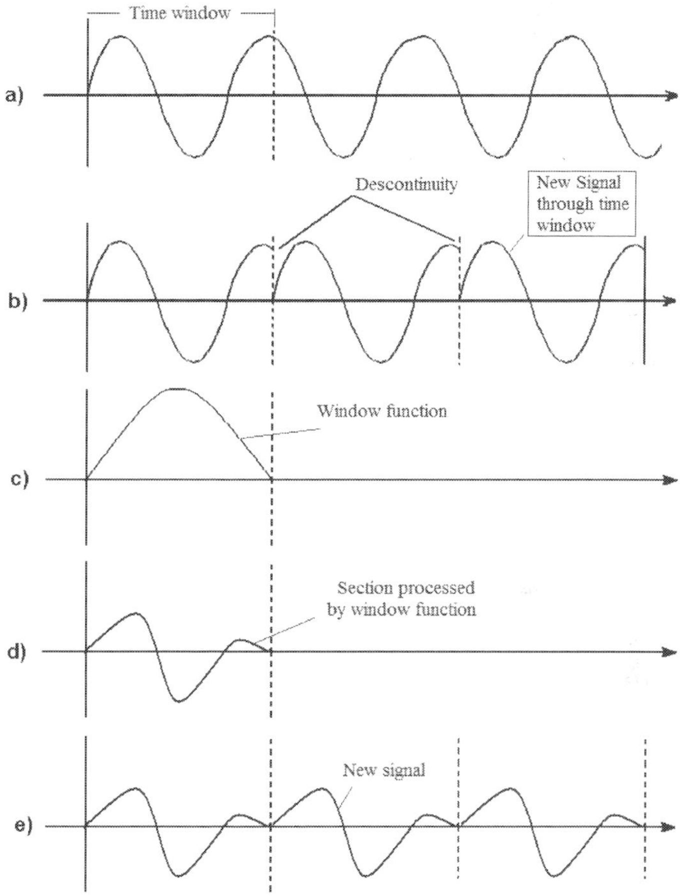

Fig. 7. Obtaining fast Fourier transform.

Depending on the structure and circumstances of the signal, some interruptions in the sequence can occur at the edges of the time window - which will reflect on the visual components of vibration.

These sequence interruptions always occur when the number of periods of the sign in the time window is not an integer. To eliminate such interruptions, a weight function is applied to the signal within the time window. As a rule, this is done so that the values of the signal at the beginning and end of the time window are reduced to zero.

Then, when the time signal on the computer is rebuilt, all the interruptions in the signal sequence are removed.

Evidently, because of this "manipulation", the machine's original vibration signal is distorted. To correct this, the results of the transformation are multiplied by a correction factor so that the exact values of magnitude are maintained, after processing.

3.3 Vibration analysis by global levels

In the case of application for predictive maintenance, the international technical standards, including the ISO, define two criteria for the adoption of a global value. One method assesses the severity of vibration by absolute measurement of non-rotating parts. The other one assesses conditions of the machine by direct measurement of oscillation of the shaft. (Gonçalves et all, 2007).

According to NBR 10082 a rating of acceptable levels of vibration severity for similar machines is established and grouped into classes. Table 3 shows the guidance offered by this standard, where:

Class I - Small machines activated by directly coupled electric motor, maximum power of 15 kW.

Class II - Mid-sized machines, class I type, with power greater than 15 kW, up to 75 kW. Motors or machines rigidly mounted up to 300 KW.

Class III - Large driving machines and other large machines (> 75 kW) with rotating masses mounted on rigid and heavy foundations, which are relatively rigid in the measuring of vibration.

Class IV - Machines of the Class III type, mounted on relatively flexible foundations in the measuring of vibration, for example, a set of turbogenerators.

Range of vibration severity		Avaliation of quality for differents machine classes			
Range	Velocity in limits (mm/s)	Class I	Class II	Class III	Class IV
0.28	0.28	A	A	A	A
0.45	0.45	A	A	A	A
0.71	0.71	A	A	A	A
1.12	1.12	B	A	A	A
1.8	1.8	B	B	A	A
2.8	2.8	C	B	B	A
4.5	4.5	C	C	B	B
7.1	7.1	D	C	C	B
11.2	11.2	D	D	C	C
18	18	D	D	D	C
28	28	D	D	D	D
45	45	D	D	D	D
71	upper 45	D	D	D	D

Table 3. Classification and assessment of machines by vibration severity levels.

Where:
A = Proper conditions;
B = Acceptable for continued operation;
C = Tolerable Limit;
D = Non-permissible.

For rotating machines with rotation speeds in the range of 600 to 12,000 rpm (10 to 200 Hz), ISO norm 2372, VDI Richiline 2056, and in Brazil by NBR 10082, take the value of effective vibration speed, known as rms speed of the signal, as the unit of measure for identifying the severity of vibration (Arato, 2004).
The parameter to be measured is the absolute velocity of vibration on the machine parts, preferably the bearings. In this case, the global value chosen as the unit of measure to indicate the vibration severity, the effective value, or simply RMS speed (V_{ef}) is not represented by a single scale of values. This is due to the great diversity of forms, mass, assembly and operational conditions of the equipment, which results in the RMS speed values for different levels of acceptable severity, (Gonçalves et all , 2007).

3.4 Demodulation

In more complex situations, where there is a combination of more than one source of excitement added to the noise transmitted through the support and foundations of the machines, the obtained spectrum of frequencies can present difficulties in the analysis (Arato, 2004).
In cases like this it is necessary to use other more dedicated techniques, such as the technique of demodulation, which enables identifying noise sources responsible for the excitation of resonant responses in the structure, hence allowing to monitor defects that are responsible for impacts of the repeated excitation type, in addition to others that produce modulator signals, even if the level of energy of the source does not allow a direct identification of its frequency in the general spectrum, as it generates amplitudes of minor significance, which remain hidden in the level of background noise.
Taking into consideration, by generalization, that the modulation in magnitude of a signal is defined as the multiplication of one sign for another, a nonlinear inherent process that creates new frequencies are not present in any of the signals involved. The identification of the noise source associated with the defect requires identifying the frequency of the modulating signal, (Arato, 2004).
The process of identifying the modulating frequency of a modulated signal is known as demodulation, and includes the following steps, (Arato, 2004):

a. Filtering of the signal by band-pass filter for the frequency range identified as modulated;
b. Detection of the modulator signal;
c. Spectral analysis of this detected modulator signal.

For the detection of the modulator signal there are several techniques. Application of the Hilbert transform that can be obtained from X (f), which is the Fourier transform of the filtered signal x(t) , according to the equations below.

$$x_{re}(t) = \mathrm{Re}\left[2\int_0^{\infty} X(f)e^{i2\pi ft}df \right] \tag{5}$$

$$x_{im}(t) = \mathrm{Im}\left[2\int_0^{\infty} X(f)e^{i2\pi ft}df \right] \tag{6}$$

Obtaining the signals $x_{re}(t)$ and $x_{im}(t)$ from which an analytical signal $z(t) = x_{re}(t) + ix_{im}(t)$ (Bendat(1986) can be constructed, (apud Arato & Silva, 2000), which can be represented by

Equation 7; where A(t) is the envelope and ϕ(t) is the instantaneous phase of the signal x(t), according to Equations 8 and 9.

$$z(t) = A(t) \cdot e^{i\phi(t)} \tag{7}$$

$$A(t) = \sqrt{x^2(t) + x_1^2(t)} \tag{8}$$

$$\phi(t) = tg^{-1}\left(\frac{x(t)}{x_1(t)}\right) \tag{9}$$

4. Materials and methods

In this work, to verify the effectiveness of the techniques studied, a reducer of the worn drive type was monitored. For this monitoring a test bench was built, where the reducer, coupled with its entry shaft to an electric motor, by means of an elastic coupling, had a load of an unbalanced mass in its output shaft.

A photograph of the bench is shown below in Figure 8. The electric motor used is a WEG, 220 V, 60 Hz, three-phase power with 0.5 CV power and 1720 rpm.

Fig. 8. Test bench for verification of the studied techniques.

The reducer used was a Macopema, ZM reducer, worn thread, with a reduction of 1:30, 0.53 CV at the entry and 0.31 CV at the output, with oil capacity of 0.25 liters.

<table>
<tr><td colspan="3">Rolling element bearing (www.skf.com)</td><td colspan="2" rowspan="2">Shaft rotation</td><td rowspan="2">Engagement</td></tr>
<tr><td>Model</td><td>6008</td><td>6204</td></tr>
<tr><td>f_{ip}</td><td>197 Hz</td><td>142 Hz</td><td>exit</td><td>entry</td><td rowspan="4">28.67 Hz</td></tr>
<tr><td>f_{ep}</td><td>147 Hz</td><td>87.5 Hz</td><td rowspan="3">0.95 Hz</td><td rowspan="3">28.67 Hz</td></tr>
<tr><td>f_{rp}</td><td>191 Hz</td><td>114 Hz</td></tr>
<tr><td>f_{gp}</td><td>13.2 Hz</td><td>11 Hz</td></tr>
</table>

Table 4. Preferred vibration frequencies of the reducer.

At the output of the reducer a bearing was attached and after the bearing a 7.5 kg mass with a 195 mm arm.
The preferred vibration frequencies of the reducer analyzed were calculated, as illustrated in Table 4.
Where:
f_{ip} = defect frequency of inner race.
f_{ep} = defect frequency of outer race.
f_{rp} = defect frequency of rolling bearing elements.
f_{gp} = defect frequency of cage.
The tests were conducted after a running period of 168 hours, for four weeks, and each week (168 hours) oil samples were collected.
Oil recommended by the manufacturer was used; beyond oil with several percentages of liquid contamination and oil with various percentages of solid contamination.
This work presents results obtained from the first four weeks of testing, with ISO 320 oil that was recommended by the manufacturer of the reducer.
The vibration measurements were collected in the three directions of the reducer. Analyses were performed in time and frequency in order to determine the beginning and severity of the active wear where the sensors were placed for collection of the vibration signs. Figure 9 shows the points along the reducer where the sensors were placed for collection of the vibration signs.

Fig. 9. Collection points of vibration signals.

According to the norms the bearings should be monitored first, thus points 3 and 7 were chosen. Points 2 and 5 represent the other two directions. These points contain all the information provided by points 1, 4, 6 and 8.
The time vibration signals were obtained by measuring the vibration speed of the reducer. For such measures piezoelectric accelerometers, a 4-channel Conditioner/ Amplifier, data acquisition system DaqBooK and a Notebook were used. The sampling frequencies were of 500 Hz, 1 kHz, 5 kHz and 10 kHz, and the corresponding analog filters were of 141 Hz, 281 Hz, 2250 Hz and 4500 kHz. For each frequency 10 samples were taken out of 2048 points.
The time vibration signals obtained were processed using the algorithm FFT (Fast Fourier Transform), and analyzed in the laboratory through the "software" DASYlab.

As the accelerometer measures the vibration speed of the reducer, using a reading indicator, the measured value of greatness is obtained directly, that is, the value of the effective vibration speed for each distinct sampling frequency. The value of the vibration severity however, is obtained when a vibration signal of a sampling frequency of 5000Hz is read, but subjected to a high-passed filter of 10 Hz and a low-passed of 1000 Hz. Both the effective value of the vibration speed and the severity of vibration were obtained using the "software" DASYlab, which contains numerous tools as: reading indicators, filters, etc.
The vibration analysis was achieved by spectral analysis, analysis by demodulation and the values of the effective vibration speed and vibration severity.
For demodulation of the signal it was necessary to use a computational routine on the Matlab platform called DEMOD, created by Arato (2004) and responsible for the calculation of Hilbert transformed.
Only the time signals obtained in points 2 and 5, when subjected to high frequencies of sampling were demodulated, due to the fact they are the only signals demonstrating frequencies that can be resonant. After demodulation the signal was processed to obtain the spectrum of the demodulated signal.
Oil samples were prepared in the rotary particle depositor (RPD), and then examined and photographed using the optical microscope Neophot 21 with adapted light transmitted. In the RPD lamina the particles are arranged in three separate rings, depending on the size of the particle, due to this, it was necessary to capture the images by observing these three rings separately.
Using the automatic monitor of ferrous particles, the PQ index of the samples was obtained. Also the viscosity, water content, the acidity index and atomic absorption of the oil samples were obtained at the end of each test.

5. Results and discussions

5.1 Analysis of lubricant

Initially, the inner elements of the reducer were photographed for a subsequent comparison and verification of wearing.

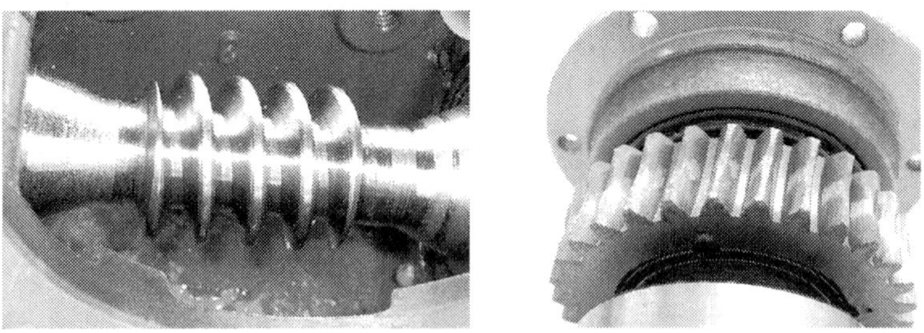

Fig. 10. Inner elements of the reducer. On the left is the worn screw and on the right the crown gear.

Figure 11 shows the photographs after the first week of the experiment.
In the figure above it can be seen severe wear particles (1^A), bronze particles (2^A), laminar particles (3^A) and cutting particles (4^A).
Figure 12 shows the photos obtained after the second week of the experiment.

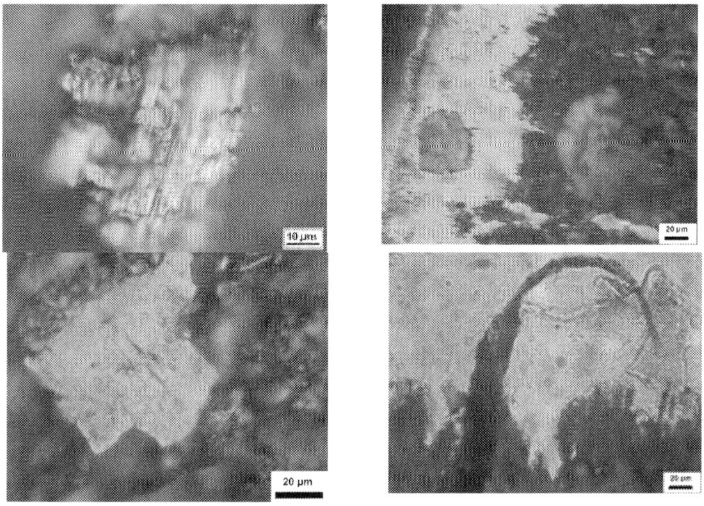

Fig. 11. Wear particles generated in the first week of the first experiment with transmitted and reflected light. First picture: inner ring of the RPD. Second photo: outer ring. Other photos: intermediary ring.

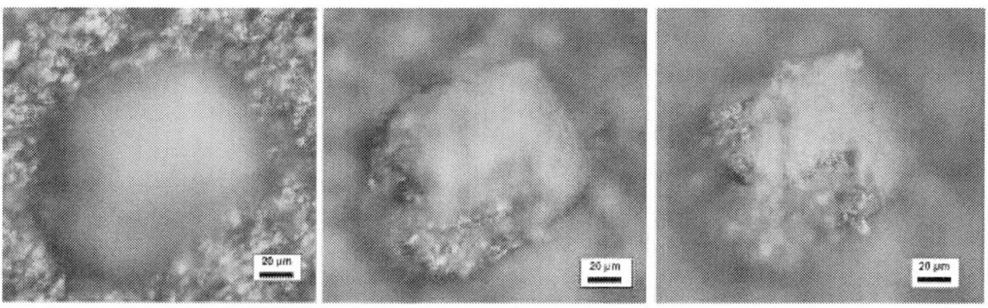

Fig. 12. Wear particles generated in the second week of the first experiment with reflected light and in the inner ring of RPD. The three photos are from the same field of vision but with different focal heights.

By Figure 12 the visual difficulty of the laminar particles can be perceived.
Figure 13 illustrates the photos obtained after the third week of the experiment.
Through Figure 14 the presence of oxide (1A), laminar particles (2A), bronze detained between the ferrous particles (3A) and parts difficult to focus can be seen (4A and 5A). The last two figures also show the need to vary the type of light in the verification of the particles.
Table 5 presents the tests values of TAN, viscosity at 40ºC and 100ºC, after the end of the last week of the first experiment. Table 6 shows the result of the atomic absorption at the end of this experiment. Table 7 shows the figures obtained in the direct ferrography during the experiments, represented by the PQ index of the magnetic particles counter. Figure 15 shows the state of the reducer after the experiments.

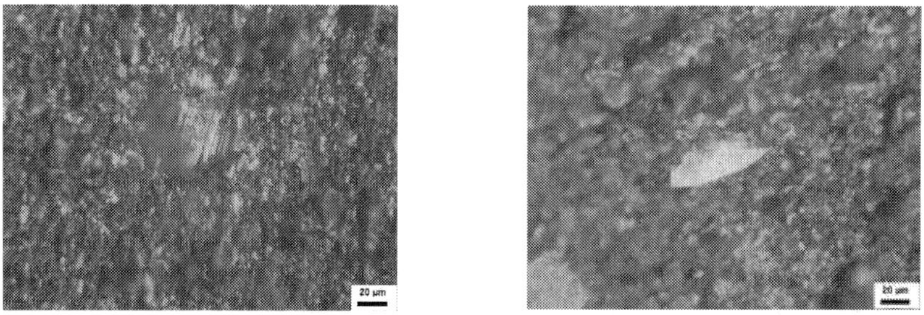

Fig. 13. Wear particles generated in the third week of the first experiment with reflected and transmitted light in the intermediary ring of RPD.

Through the figure above the severe wear particles represented by striation (1^{A}) and the particles overlapping the others (2^{A}) can be seen.
In Figure 14 the photos obtained after the fourth week of the experiment are shown.

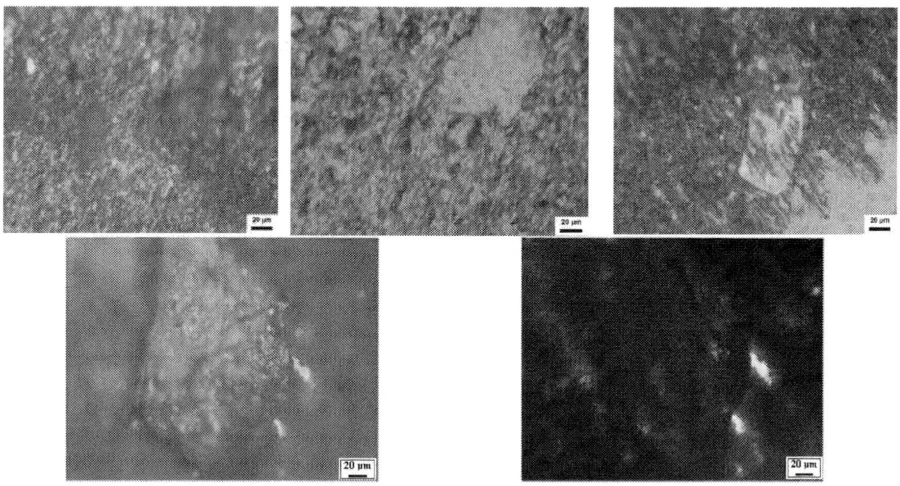

Fig. 14. Wear particles generated in the fourth week of the first experiment The first photo is of the inner ring of RPD observed with transmitted and reflected light. The second photo is of the outer ring with the same lights. The third one is the intermediary ring with the same lights. The last two are also of the intermediary ring; the second to last is with transmitted and reflected light and the last one with only transmitted light.

Water	TAN	Viscosity at 40ºC	Viscosity at 100ºC
ASTM D 95	ASTM D 664	ASTM D445	ASTM D 445
(%)	mg KOH/g	cSt	cSt
0.00	1.17	295	42

Table 5. Tests conducted on the oil at the end of the experiment.

Atomic absorption						
Cu	Si	Al	Fe	Pb	Cr	Ni
Ppm	ppm	ppm	ppm	ppm	ppm	ppm
580	18	4	142	0	0	7

Table 6. Atomic Absorption conducted on the oil at the end of the experiment. Atomic absorption

Test A			
1st Sample	**2nd Sample**	**3nd Sample**	**4nd Sample**
670	1680	3050	4000

Table 7. PQ Rates obtained on the particle monitor.

Fig. 15. Inner elements of the reducer after the first test.

Through Table 5 it was noticed that the viscosity at 100 ºC increased from allowable range (27.9-33.3), according to the specifications of the new oil, to 42.03. This means an alert since the 10% of the permissible range was exceeded. This may be an indication of oxidation of the oil.

In Table 6 large amounts of Cu was noticed in the bronze of the crown and Fe in the alloy steel from which the pinion is manufactured. The Si is an indicator of external contamination.

In Table 7 the gradual wear of the pinion can be seen. As the samples were not changed during the first experiment, the quantity of metallic particles accumulated during the weeks of this first experiment.

5.2 Analysis of vibrations

Several vibration measures were taken at various points of the reducer. In the following Figures some measures taken at some points are presented. Table 8 shows the effective value and the severity value of the vibration velocities.

	Vibration effective value (mm/s)			**Vibration Severity**
	500Hz	5000Hz	10KHz	NBR 10082
Point 2	0.27	0.71	0.99	0.64
Point 3	0.34	0.43	0.54	0.51
Point 5	0.37	0.44	1.09	0.42
Point 7	0.51	0.64	0.68	0.61

Table 8. Effective value of vibration speeds (mm/ s) and severity values of vibration by the NBR 10082 norm, (10 to 1000Hz), at the end of the last week of the first experiment.

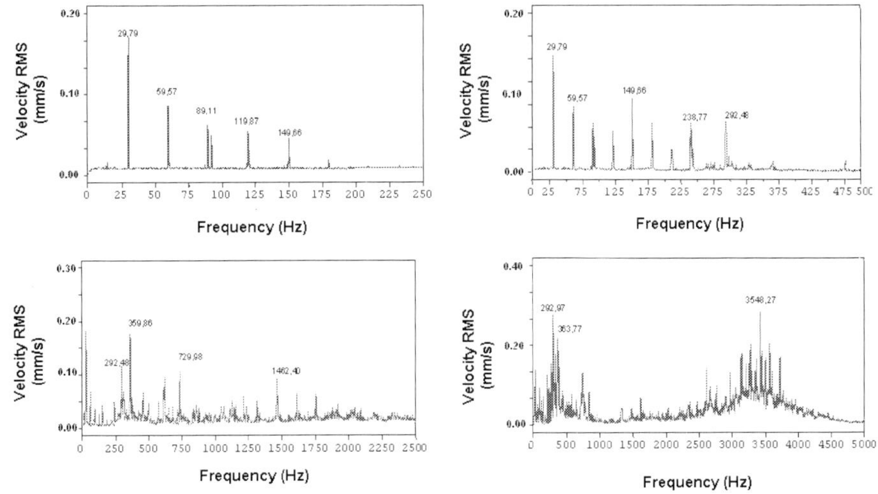

Fig. 16. Vibration spectra obtained at point 2, for sampling frequencies of 500, 1000, 5000 and 10000 Hz with analog filters of 141, 281, 2250 and 4500 Hz respectively, at the last week of the first experiment.

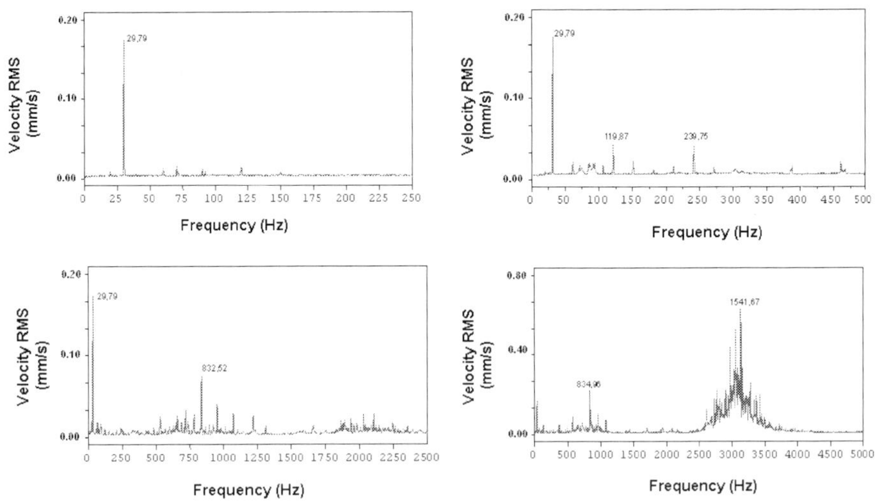

Fig. 17. Vibration spectra obtained at point 5, for sampling frequencies of 500, 1000, 5000 and 10000 Hz with analog filters of 141, 281, 2250 and 4500 Hz, respectively, after the last week of the first experiment.

Figure 18 presents a graph showing the trend of the values obtained in the four weeks of the first test for points 2, 3, 5 and 7.

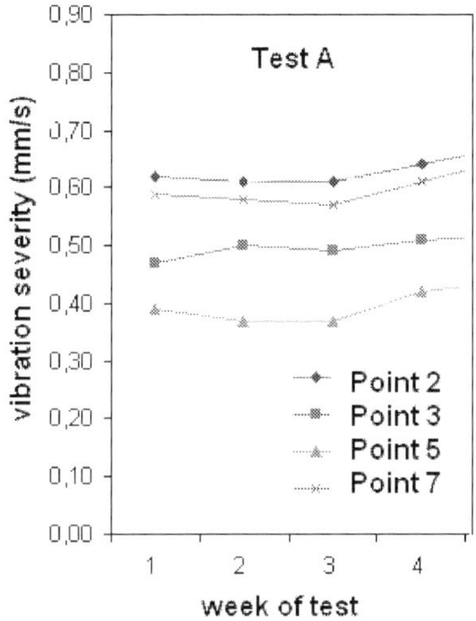

Fig. 18. Evolution of the vibration severity in accordance with the NBR 10082 norm.

Considering that our experiment fits in Class I of the NBR 10082 norm, we found that all the values are within the range considered to be in good conditions, that is, within the range A of Table 3.
Resonant frequencies were found in points 2 and 5 of the reducer. These were demodulated when subjected to a sampling frequency of 10kHz.
Figure 19 shows Point 2 with a cutoff frequency of 250 and 450Hz, where demodulation was performed on the sampling signal of 10 KHz at the end of the first experiment.

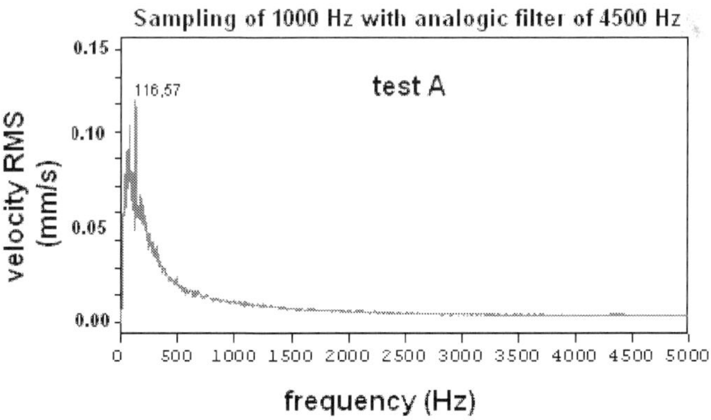

Fig. 19. Filtered time signal around the resonance frequency using a cutoff frequency of 250 Hz and 450 after demodulation by the program DEMOD followed by its spectrum.

Figure 20 shows Point 5 with a cutoff frequency of 2900 and 3300Hz, where demodulation was done on the sampling signal of 10 KHz at the end of the first experiment.

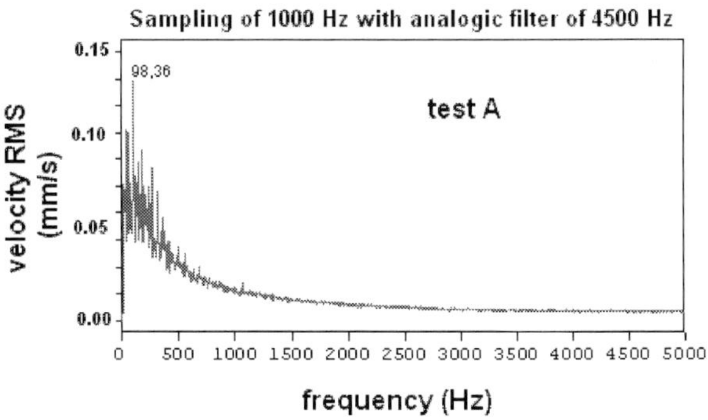

Fig. 20. Filtered time signal around the resonance frequency using a cutoff frequencies of 2900 and 3300 Hz, after demodulation by the program DEMOD followed by its spectrum.

Through demodulation of the signal it was observed that these frequencies are not related to bearing defects.

6. Conclusions

The type of reducer used, worn drive, is difficult to analyze because in this case there is a low rotation shaft (exit of the crown, which in our case is of 0.95 Hz) and the engagement frequency coincides with the rotation shaft frequency, given that the worn screw has a single entry.

In the studied reducer the oil analysis showed a certain advantage, given that since from the beginning this operation demonstrated the reducer's improper operation, while the vibration analysis was difficult to use given that the rotation shaft and engagement coincided.

The experiments also demonstrated that if in a system an abnormal occurrence takes place, for example an extra load during a certain period of time, the subsequent vibration analysis may not perceive what took place if no permanent damage occurred, however the oil analysis will acknowledge that there was such a problem during the operation.

No bearing defects were found in the spectra and neither in the demodulation.

After disassembling the reducer it was found that the axis bearings of the worn screw were loose. The literature shows that the loose bearings are represented by the frequency of the spin axis and its harmonics, which is the same case as the wear. This fact was difficult to check because of the coincidence of such frequencies.

One of the simplest tests, the counting of particles through the automatic monitor of PQA particles, showed a high generation of magnetic particles from the first sample.

The atomic absorption performed at the end of each test also provided important information about the tribological condition of the system, where a large amount of copper particles can be noted from the crown and iron resulting from the pinion.

The reducer's incorrect operation can also be verified by a visual inspection of the particles. Most of the photos demonstrate the presence of many ferrous particles and some copper particles. We also found many wear particles by friction (laminar particles) and severe wear particles by sliding, which are particles with striation.

7. Acknowledgment

For Continuos Funding and support, I thank:
FUNDUNESP - Unesp Development Foundation;
FAPESP- State of Sao Paulo Research Foundation;
CNPq - National Counciul for Scientific and Technological Development;
FINEP - National Fund for Studies and Projects.

8. References

Anderson A, Sweeney A, Willians G.(1999). How to make decisions. In: ___ Quantitative approaches to decision making. 9.ed. South Western: College Publishing, 1999. p.666-671.

Arato, A.(2004). Manutenção preditiva usando análise de vibrações. Barueri: Manole, 2004. 190 p.

ASSOCIAÇÃO BRASILEIRA DE NORMAS TÉCNICAS - ABNT. NBR 10082: vibração mecânica de máquinas com velocidades de operação de (600 a 12000) rpm - bases para especificação e padrões de avaliação, 1987.

Azovtev, A.; Barkov, A.(1998). "Development of Condition Diagnostics in Russia Using Vibration", Vibro Acoustical Systems and Technologies, Inc. (VAST), Saint Petersburg, Russsia, 16p., 1998.

Barraclough T G, Sperring T P, Roylance, B J, Nowell, T.(1999). Generic-based wear debris identification -on the first step towards morphological classification. In: PROCEEDING OF THE INTERNATIONAL CONFERENCE ON CONDITION MONITORING, 1999, Swansea. Proceedings... Oxford: Coxmore Publishing, 1999. p.525-538.

Bendat, Julio & Pierson, Allan G.(1986). "Randon data: Analysis and Measurment Procedures", John Wiley, EUA, 1986.

Diana, G.; Chelif, F.(1999). "Vibration Mechanics", SFM: Société Française des Mécaniciens, Acoustical and Vibratory Surveillance Methods and Diagnostic Techniques, Senlis, France, p.475-486, 10-12 October, 1999, vol 2/2.

Gonçalves A C, Lago D F, Cunha R C. (2007). Vibration and wear Particles Analysis in a test Stand. Industrial Lubrication and Tribology. Vol 59, issue 5, 2007, ISSN: 0036-8792

Gonçalves, A. C. ; Campos-Silva, J.B.(2011). Predictive Maintenance of a Reducer with Contaminated oil Under an Excentrical Load Throug Vibration and Oil Analysis. Journal of the Brazilian Society of Mechanical Sciences and Engineering (Impresso), v. XXXIII, p. 1/1-7, 2011.

Gonçalves, A. C. ; Padovese, L. R.(2010) . Vibration and oil analysis for monitoring problems related to water contamination in rolling bearing. In: First International Brazilian Conference on Tribology - TriboBR 2010, 2010, Rio de Janeiro. First International Brazilian Conference on Tribology - TriboBR 2010 and ITS - IFTo MM 2010 - 2nd International Tribology Symposium of IFToMM, 2010.

Grenn A. (2003). "The study of the failure modes of a spur gear using vibration an particle analysis techniques" JAMES COOK UNIVERSITY School of engineering, 2003

Hutchings, I. M.(1992). Tribology: friction and wear of engineering materials. Londres: Edward, 1992.

Kato K.; Adachi K. Wear mechanisms.(2001) In: BHUSHAN, B. (Ed.). Modern tribology handbook: principles of tribology. Boca Raton: CRC Press, 2001.

Kimura, R. K. ; Gonçalves, A. C.(2009) . Análysis of biodiesel type and contamination level on lubricant properties of diesel stationary enginees. In: 20th International Congress of Mechanical Engineering, 2009, Gramado. 20th International Congress of Mechanical Engineering -COBEM2009, 2009.

Lockwood F E, Dalley R. Lubricant analysis. (1992). In: HENRY, S. D. et al. (Ed.). ASM Handbook: friction, lubrication, and wear technology. Metals Park: The Materials Information Society, 1992. v.18, p.299-312.

Mirshawha, V. (1991). "Manutenção Preditiva: Caminho para Zero Defeitos", São Paulo, Makron, McGraw-Hill, 1991, 317p.

Modern Tribology Handbook, Vol. I ("Principles of Tribology"), Edited by Bharat Bhushan, CRC Press, 2001

Parker.(2000). The handbook of hydraulic filtration. Metamora: Parker, 2000. p 12-19.

POA. (2002). "Practicing Oil Analysis", November - December, 2002, p. 46 - 50.

Rabinowicz, E.(1995). Friction and wear of materials. 2. ed. New York: John Willey and Sons, 1995.

Wang, H.; Williams, K.(1995). "The Vibrational Analysis and Experimental Verification of a Plane Electrical Machine Stator Model", Mechanical Systems and Signal Processing, p.429- 438, 1995.

6

Probabilistic Vibration Models in the Diagnosis of Power Transformers

Pablo H. Ibargüengoytia[1], Roberto Liñan[1], Alberth Pascacio[1]
and Enrique Betancourt[2]
[1]*Instituto de Investigaciones Electricas*
[2]*Prolec General Electric*
Mexico

1. Introduction

Power transformers are some of the most important equipment for the transmission and distribution of electric power. A single failure in a transformer causes disturbances in the electric network and may cause severe conflicts in hospitals, banks, industrial installations or urban areas in general.

In Mexico, the transmission network is composed by 350 power substations and 2,580 power transformers. The capacities of these transformers are typically 375, 225 and 100 MegaVoltAmpere (MVAS), with a nominal tension of 400 kV, 230 kV and lower. Approximately 27% of these transformers have more than 30 years in operation. For this reason, it is important to observe and register the amount and type of failures that have presented the transformers in the country.

Table 1 shows the type of transformer failures from 1997 to 2007 (CFE, 2010).

Failure	1997	1998	1999	2000	2001	2002	2003	2004	2005	2006	2007	Total
winding insulation	11	6	5	10	5	2	9	4	6	5	6	69
core	0	0	0	0	0	0	0	0	0	0	2	2
bushing	3	2	5	3	1	3	1	1	5	7	5	36
on load tap changer	2	0	2	2	2	2	1	1	2	3	1	18
explosion with fire	1	3	0	0	2	0	2	0	3	0	0	11
other failures	1	0	1	0	2	4	1	2	0	0	0	11
TOTAL	18	11	13	15	12	11	14	8	16	15	14	147

Table 1. Type of failures in transformers since 1997.

Notice that the highest percentage of failures corresponds to isolation in the windings. Problems with insulation represent 80% of the failures for contamination, aging and core insulation. Other causes can be overvoltage or short circuits. The insulation failures can be slow degradation, while overvoltage and short circuits represent instantaneous failures. The Mexican case is not unique. Failures in transformers in United States and Russia show similar results.

All these common failures can be considered as mechanical failures, in contrast to other chemical based failures. Consequently, mechanical failures cause vibration.

Literature reports different methods for the diagnosis of power transformers. Among the most important are the following:

- dissolved gas analysis: analyzes the chemical composition of the isolating oil. Percentages of different gases are found and tables provide a relation between these percentages and the status of the transformer. It is widely used to detect all the faults related with oil.
- frequency response analysis: analyzes the transformer response given a voltage or current excitation. The response is analyzed in the domain of frequency in magnitude and phase. Later, this response is compared with known reference responses and then, a difference from the correct operation can be detected. Also, a signature of certain faults can be obtained and compared.
- partial discharge: utilizes acoustic sensors to detect the location of partial discharges occurrence inside the transformer. A partial discharge is a transient electric discharge that bridges the isolation gap between two conductors. It is then possible to detect deficiencies in the transformer isolation.

An alternative method for detecting failures in transformers is the analysis of the vibration produced inside the transformer due to its operation. Normally, the transformer produces vibrations in the windings and the core, and these vibrations vary according to certain operative conditions. Also, in the presence of mechanical failures, the vibration pattern is different than in normal conditions. One advantage of this alternative method for detecting failures, with respect to the mentioned methods is that it is possible to design an on-line diagnosis system. This implies that the detection of the incipient failures can be achieved at all times, while the transformer is working.

This chapter presents the development of a probabilistic model of the vibrations in a transformer given all the possible combinations of operating conditions. The probabilistic models are Bayesian networks (BN). The BN are directed acyclic graphs that represent the probabilistic relation between the variables in a domain. In this case, we obtain a Bayesian network that codifies the relation between the vibration signals with all the variables that conform the operative conditions. The model is obtained using automatic learning algorithms applied to historical data of the transformer working at different conditions. The BN is used to calculate the probability of a failure, given the evidence from the operative conditions.

The next section establishes the central problem in this project, namely the vibrations in a power transformer. Also, section 2 reviews reviews some of the related work reported in the literature.

2. Power transformers

The transformer is an electrical device without moving parts, that is used to transfer energy from one circuit to another through a common magnetic field, with no direct electrical connection between the two circuits, (Harlow, 2007).

Basically consists of two or more windings (coils) and a common core of steel. When a voltage is applied in the exciting coil, the coils exchange energy via magnetic induction that occurs when an electric current variation in time is passed in one of them.

Figure 1 shows the idealized model of a two windings transformer with steel core. When applied an alternating voltage ($V1$) to the exciting winding, a current ($I1$) flows and generates a magnetic flux (ϕ). The flux circulates in the steel core so that the voltage ($V2$) is induced in the second winding. The ratio of applied voltage and induced voltage ideally depends on the number of turns of each coil. For example, if the number of turns in the primary winding is twice that of the secondary winding, then the voltage $V2$ induced in the secondary is half of $V1$. This considering that almost 100% of the flow is driven by the core. In power transformers is commonly used steel laminated core for high capacity that has to drive the magnetic flux (steel core laminated from very thin sheets, such as .23 mm for reduced *eddy* and *hysteresis* losses). The ability of a material to conduct magnetic flux is called permeability. Currently, there are electrical steels for power applications with permeability in the order of 1,500 compared with 1.0 for air.

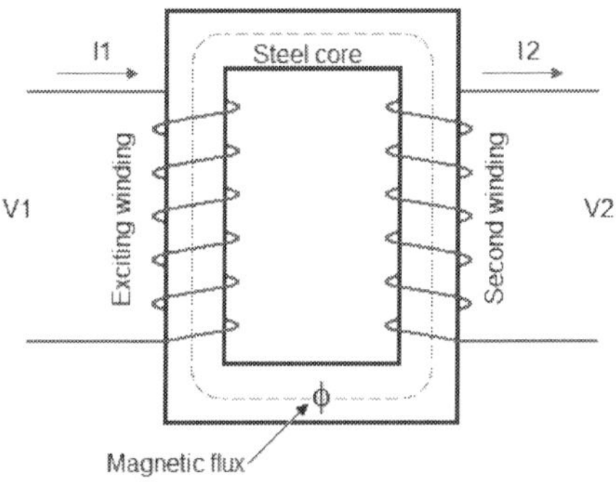

Fig. 1. Ideal transformer: two coils applied on steel core.

Transformers can be classified according to their application, construction or power capabilities. In the electricity transmission process, the utilized power transformers are characterized by the high levels of voltage and current that they handle. Power transmission requires power transformers to transmit energy from voltage production centers over long distances to consumption areas. Here, power transformers decrement the voltage down again for distribution to consumers. The distribution process requires transformers with less voltage and current. They are called distribution transformers.

For the design and construction of power transformers, regardless of size and power, various materials are used such as:

- Thin magnetic steel plates in the core.
- Copper or aluminum as conductor material in the coils (see Fig. 2).
- Cellulose products like high density paper and pressboard as solid insulation material.
- Mineral oil as liquid insulating material, which also functions as a coolant.
- Steel for the tank.

Additionally, the transformer has several external accessories such as (see Fig. 3):

- Radiators, fans and oil circulating pumps for the cooling function.
- Bushings as an insulating structure, for connection of windings to the outside.
- Oil temperature indicators.
- Oil Level Indicator

Fig. 2. Inner part of the transformer. The core and coils assembly is shown at the manufacturing stage. (Courtesy of Prolec GE).

- Pressure relief devices.
- Winding temperature indicator.
- Sudden pressure relay.
- Desiccant breathers.
- Liquid preservation systems.

2.1 Vibration in transformers

Transformers always vibrate while operating. Vibration can be detected at different frequencies, in different places of the transformer and caused by different sources. According to the literature, vibration below 100 Hz. is caused by cooling fans and oil pumps. Vibrations above 1000 Hz. are caused by small elements not related to the state of the core or winding (Golubev et al., 1999). Thus, the vibration frequency range of interest in transformer diagnosis is between the power frequency of 60 Hz (50 in Europe) and multiples of this up to 960 Hz.

In normal operating conditions, the main sources of vibration are the core and the winding of the power transformers. This vibration is transmitted to the transformer tank through the

Fig. 3. Power transformer at the electrical substation. (Courtesy of Prolec GE).

cooling oil and through the solid structure. Different levels of vibration can be measured at different locations of the transformer.

Vibration in the winding is caused by Lorenz forces that depend on the current density and the leakage flux density. Since the leakage flux and the current density have different directions, the winding force density has components in the radial and axial directions. Also, both components are a function of loading current, so the total Lorenz forces are quadratic functions of the current. It is worth to mention that the vibration caused by the winding is not too significant under normal operating conditions, but it is significant under several kinds of failures.

Vibrations at the core are caused by the magnetostriction process. It consists of changes of the dimension of core laminations, made by ferromagnetic materials, due to changes of orientation of the material crystals for magnetic fields. Thus, magnetostrictive effects are function of the magnitude of the applied field. Also, it is known that these magnetostrictive forces have a fundamental frequency of 120 Hz. i.e., twice the exciting frequency. Other source of core vibration is the air gap produced by the magnetic repulsion among laminations. This repulsion forces are mostly present at the corner joints of the core legs with the jokes, and it has also a fundamental frequency of 120 Hz (Lavalle, 1986).

In case of a failure like a short circuit, the mechanical integrity of the transformer can be altered. Certain changes, such as loss of winding clamping pressure, will lead to insulation

deterioration, and therefore, the vibrational response will be altered. In general, most of the failures occurring in the transformer, produce mechanical deformations in the winding, and hence, a change in the vibrational signature of the transformer. Typical failures occurring at the transformer core are caused by short circuits between core laminations or between the core and the tank. Since the tank is grounded, if several contact points between core and tank exist, a current will flow and temperature will rise. Also with laminations short circuits, the temperature increases. In general, temperature changes will produce changes in the insulation system that will produce mechanical changes that provoke changes in the vibrational signature of the transformer.
Summarizing, most of the failures in the winding and core of the transformer, will produce mechanical changes and will turn out in changes in the vibration pattern of the transformers.

2.2 Related work in transformers diagnosis by vibrations

Several technical groups have worked in the study of vibrations for identifying and diagnosing mechanical faults. The basic approach followed by all the research groups is the development of a model that represents the behavior of the transformer under different conditions. Later, when new readings are obtained, the model estimates or predicts the behavior of the transformer and a comparison with the current behavior can identify abnormal events. The key question here is *how can this model be defined?*. This question has been answered differently by some research groups.
The Electrical Engineering and Computer Science Department at the Massachusetts Institute of Technology (MIT) elaborated a series of Master of Science thesis devoted to the development of vibration models for faults identification in transformers (Crowley, 1990; Lavalle, 1986; McCarthy, 1987). First, they utilized an analytical model that relates the harmonic content of one input (square of the current) to one output (the vibration in the winding). The basic model has the form $y = Cu$ where y is a vector whose components are the amplitude of the vibrations discomposed at frequencies multiple of 120Hz. u is the current squared vector at the some frequencies and C is the matrix with parameters that relate them. The elements c_{ij} of C are estimated in a set of experiments made to a transformer free of faults at different conditions of load. For a given load condition, and for each frequency, several measurements were averaged and related with a Least-Squares algorithm to define the c_{ij} parameters. Other experiments were made to relate effects of the temperature in the same structure of measures. A third set of experiments was carried out in presence of a fault: a loosened up the windings. Other experiments consider the effects of the core vibrations in the winding vibrations. When both the winding and the core are excited simultaneously, the sources of vibration may have phase differences. They proposed a model that includes linear compensation for winding temperature, weighted by the harmonic amplitudes of the loading current squared, and linear compensation terms for core temperature and excitation voltage. Summarizing, the MIT theses propose a model that requires the acquisition of certain parameters from experiments implemented off-line. Later, these parameters are compared with the on-line parameters to detect changes in the normal behavior of the transformer.
Spanish Unión Fenosa proposes a model that is also based on vibrations (García et al., 2006a;b). They measure vibrations at the tank of the transformer, claiming that the vibrations are transmitted to the tank from two sources: the winding and the core. Thus, the vibration at the tank results from the vibration of the windings plus the vibration of the core, multiplied by a coefficient of transmission. The winding vibration is proportional to the square of the

current, and the vibration at the core is proportional to the square of the voltage. They also consider the temperature of the transformer as an important parameter in their model, so they complemented their analytical model with complex variables that represent the real and imaginary part of the amplitude of the vibration, the current and the voltage respectively, at the main frequency component. Other parameters are the oil temperature, and the geography of the transformer. These parameters must be defined through measurements taken off-line for each kind of transformer. Their diagnosis method consists of the estimation of the tank vibration and its comparison with the real measure. If the difference is greater that certain threshold, then a fault is detected.

The Russian experiments (Golubev et al., 1999) install accelerometers in both sides of the transformer in order to acquire vibration measurements while the transformer is working properly. They executed two sets of experiments. In the first experiments, no load is included in order to detect the vibration pattern due to the core. In the second set of experiments, load is included for detecting vibration from both, core and winding. Thus, they subtract the effect of both minus the effect of the core to deduce the effects of the winding. With this information, they calculate four coefficients that reflect the clamping pressures. If these coefficients exceed 90%, then the clamping pressure is in a good state. Between 80% and 90%, the pressure is in a fair state but the transformer can continue operating. Below 80%, the pressure is critical and requires immediate attention. This approach has been tested in more than 200 transformers 110-500 kV to 50 MVA in Russia with a rate of more than 80% confirmed diagnosis. Also, Manitoba Hydro power plants in Canada tested their large power transformers with this methodology with good results.

The approaches commented above, and our approach have similar basis. All utilize vibration measures in the tank of the transformer. All transform the vibration signals to the frequency domain in order to process the vibration components at the different frequencies. All propose a model that is utilized to estimate vibration amplitude values, and then compare with real measurements in order to detect changes in the behavior. In the revised work, models are deduced with analytical equations to define certain parameters that have to be acquired off-line over a testing transformer. Experiments are required over different operating conditions and also, in presence or absence of different faults. All these approaches deduce a general model for all kind of transformers where the experiments define the specific parameter for each kind of transformer.

The approach proposed in this chapter also utilizes a model. However, this model represents the probabilistic relations between condition operational variables and vibration measurements. This implies some special advantages:

- several automatic learning algorithms are available for model construction,
- empirical human expertise can be included in the models,
- the models can be adapted constantly for each kind of transformer in its real operational condition. This means that the diagnosis may still work even if the transformer is old and vibrates more that when new, but still working properly.
- other sources of information can be included, for example, structural characteristics of a transformer.

The next section describes basis for the proposed model.

3. Probabilistic modeling

The basic idea in this work is the representation of the vibration behavior of the transformer under different operational conditions. This allows detecting deviations of the normal behavior of the transformer. Therefore, the idea is to calculate the probability of an abnormal behavior, given the operational conditions and the vibration measured. The representation of the behavior is built using probabilistic models and specifically Bayesian networks.
The basic idea is the following. Calculating the probability of an abnormal behavior (hypothesis H) can be made using the evidence recollected (E) and the Bayes theorem as follows:

$$P(H \mid E) = \frac{P(E \mid H)P(H)}{P(E)} \tag{1}$$

For example, if we want to calculate the probability of a windings loosened up hypothesis ($P(H \mid E)$) given that we observe high vibration as evidence, we could easily calculate by counting the times that we observe high vibration given that we knew that the transformer has loosened up windings ($P(E \mid H)$). However, if multiple hypotheses exist, and multiple evidence can be obtained, then the Bayes theorem in this form is not practical. What is needed is a practical representation of the dependencies and independences between the variables in an application. This representation is formed by the Bayesian networks (BN).
Formally, a Bayesian network is defined as a directed acyclic graph, whose nodes represent the variables in the application, and the arcs represent the probabilistic dependency of the connected nodes (Pearl, 1988). The Bayesian network represents the joint probability distribution of all variables in the domain. The topology of the network gives direct information about the dependency relationship between the variables involved.
As an example, assume that some application deals with the following variables: temperature (*temp*), excitation with voltage (*voltage*), load (*load*), amplitude of the acceleration (*amplitude*) and frequency (*freq*). Suppose for this example that voltage excitation of the transformer produces an increase of the temperature and a variation on the load fed. Also, the load produces an increment on the acceleration and variations of the frequency of this acceleration. This knowledge can be represented in a Bayesian network as shown in Fig.4. In this case, the arcs represent a relation of causality between the source and the destination of the arcs, according to the text above. Variables *load* and *temperature* are probabilistically dependent of variable *voltage*. Also, variables *frequency* and *amplitude* are dependent on *load*. Notice that besides the representation of the dependencies, the representation of the independences is an important concept in BN. In this example, *frequency* is probabilistically independent of *voltage* given *load*. Also, *amplitude* is independent of *temperature*.
Using the dependency information represented in the network, and applying the chain rule, the joint probability function of the set of variables in the application is given by:

$$P(t,l,v,f,a) = P(freq \mid load)P(amplitude \mid load)P(load \mid voltage)P(temp \mid voltage)P(voltage)$$

This corresponds to the product of $P(node_i \mid parents(node_i))$.
Besides the knowledge represented in the structure, i.e., dependencies and independencies, some quantitative knowledge is required. This knowledge corresponds to the conditional probability tables (CPT) of each node given its parent (corresponding to the term $P(E \mid H)$ in the Bayes theorem) and a-priori probability for the root nodes (corresponding to the term $P(H)$ in the Bayes theorem).

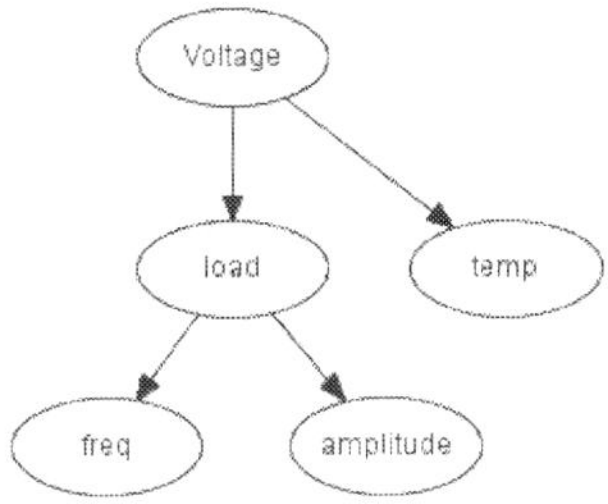

Fig. 4. Example of a Bayesian network with 5 variables.

Thus, a complete probabilistic model using Bayesian networks is formed by the structure of the network, and the CPT tables corresponding to each arc, and a-priori vectors corresponding to the root nodes (nodes without parent).
One of the advantages of using Bayesian networks is the three forms to acquire the required knowledge. First, with the participation of human experts in the domain, who can explain the dependencies and independencies between the variables and also may suggest the conditional probabilities. Second, with a great variety of automatic learning algorithms that utilize historical data to provide the structure, and the conditional probabilities corresponding to the process where data was obtained (Neapolitan, 2004). Third, with a combination of the previous two, i.e., using an automatic learning algorithm that allows the participation of human experts in the definition of the structure.
Once that the probabilistic model has been constructed, it can be used to calculate the probability of some variables given some other input variables. This consists of assigning a value to the input variables, and propagating their effect through the network to update the probability of the hypotheses variables. The updating of the certainty measures is consistent with probability theory, based on the application of Bayesian calculus and the dependencies represented in the network.
For example, in the network in Fig. 4, if *load* and *temp* are measured and *freq* is unknown, their effect can be propagated to obtain the posterior probability of *freq* given *temp* and *load*.
Several algorithms have been proposed for this probability propagation. For singly connected networks, i.e., networks in what all nodes have at most one parent as in Fig. 4, there is an efficient algorithm for probability propagation (Pearl, 1988). It consists on propagating the effects of the known variables through the links, and combining them in each unknown variable. This can be done by local operations and a message passing mechanism, in a time that is linearly proportional to the diameter of the network. The most complete and expressive Bayesian network representation is multiply connected networks. For these networks, there are alternative techniques for probability propagation, such as clustering, conditioning, and stochastic simulation (Pearl, 1988).
This project obtains historical data from different accelerometers collocated in different parts of the prototype transformer. The transformer is operated at different conditions of load, temperature, and excitation. The data acquired is fed to an automatic learning algorithm that produces a probabilistic model of the vibrations in the transformer working under different conditions. Thus, given new readings in a testing transformer, the model calculates through probabilistic propagation, the probability of certain vibration amplitudes at certain

frequencies. Therefore, a deviation of this behavior can be detected when reading the current values of acceleration and frequency. The next section explains this process detailed.

4. Probabilistic vibration models

Two approaches were considered for the diagnosis of transformers based on vibration signals. The first approach consists of inserting failures in a transformer and measures the vibration pattern according to the operational conditions. The diagnosis becomes a pattern recognition procedure according to the set of failures registered. Some examples of common failures are loosening the core or loosening the windings. These failures are similar to those failures caused by strikes or short circuits. The second approach consists of the measurement of vibration signals of a correct transformer working at different operational conditions. These measures allow the creation of a vibrational pattern of the transformer working properly. Only one model is obtained in this approach. Only measures in a correct transformer are required. As a consequence, this second approach is reported in this chapter, i.e., the construction of a model for the correct transformer.

Additionally, two sets of experiments were conducted. In the first, experiments considered the operational tests performed at the factory in the last steps of the construction of the transformers. These tests increments the number of factory acceptance tests (FAT). The second set of experiments considers the normal operational conditions of the transformer and detects abnormal behavior in site (SAT).

In the next section, we include a description of the experiments conducted, and the construction of the model of correct transformer. Finally, we discuss the difference between FAT and SAT models.

4.1 Experiments

The creation of a model for the correct functioning of the transformer requires correct transformers. The experiments were done at the Prolec-General Electric transformer factory in Monterrey, Mexico. We had access to the production line at the last step of the new transformers tests. We installed 8 sensors around the transformer as shown in Fig. 5: two in each side, one in the lower and the other in the upper part of every side. This array of sensors permits us to identify the specific points of the transformer where the vibrations signals can be detected properly.

Experiments in Prolec GE factory consisted in 19 different types of operational conditions. Table 2 shows the operational conditions and the effect we wanted to study.

	Temperature		
Excitation	Cold	Hot	Condition
Voltage	Effect of voltage in core vibrations	Effect of voltage and temperature in core vibrations	70%, 80%, 90%, 100%, 110%
Current	Effect of current in winding package vibrations	Effect of current and temperature in winding package vibrations	30%, 60%, 100%, 120%

Table 2. Type of experiments in factory.

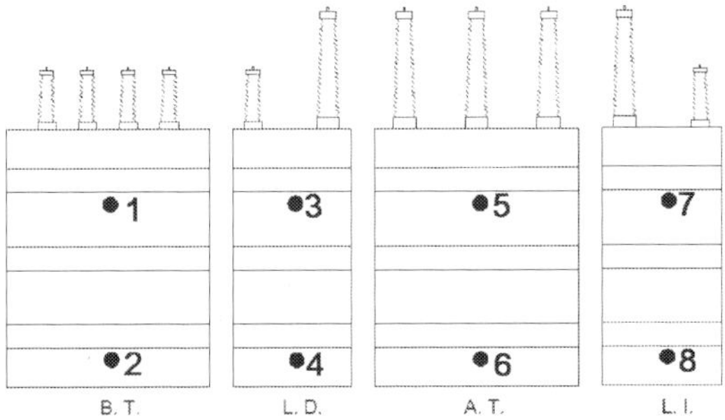

Fig. 5. Location of the sensors in the transformer. Two in the low voltage side (B.T.), the following in the right side (L.D), two in high voltage side (A.T) and the last in the left side (L.I).

Fig. 6. Transformer in Prolec GE factory with the sensors (Courtesy of Prolec GE).

The experiments combine temperature and excitation. The experiments with cold transformer excited with voltage and no current are used to study the effects of voltage in core vibrations. Cold transformers excited with current and no voltage are used to study the effects of current in winding packages. Hot transformers with voltage study effects of temperature and vibration in the core. Finally, hot transformers and current study the effects of temperature and vibrations in the winding. Additionally, the experiments that study the effects when excited with current and no voltage, included variations between 30%, 60%, 100% and 120% of the nominal current for each transformer. Every transformer report its nominal current and nominal voltage. Similarly, the effects when excited with voltage and no current included variations between 70%, 80%, 90%, 100% and 120% of the nominal voltage. In total, 19 different types of experiments were conducted to all the transformers.

For each experiment, once that the transformer is prepared to a specific test, our data acquisition system collects vibration data at 5 K samples per second during two seconds for each sensor. Later, we apply the discrete Fourier Transform (DFT) and extracts the frequency content of the data set acquired. This is repeated ten to twelve times for each operational condition.

Repeating this procedure for all operational conditions, for all the sensors, we obtain the graphs as shown in Figures 7 to 10. Notice that the only information that we need to extract with the DFT is the frequency content of the vibration at frequencies multiple of 60Hz. In fact, we find no other components in frequencies different than these multiples.

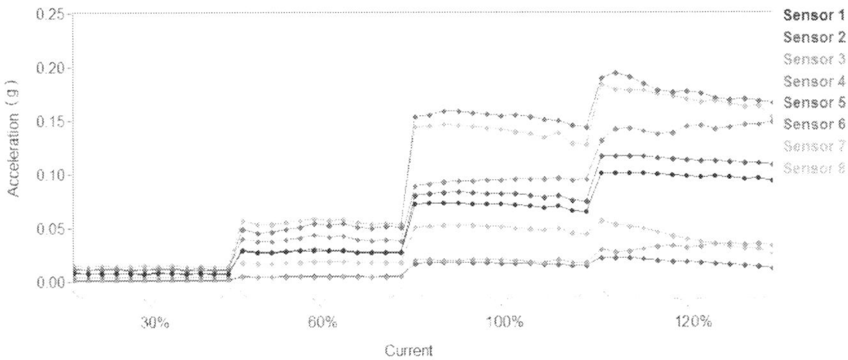

Fig. 7. Vibration signals when excited with current at 120 Hz. in all sensors.

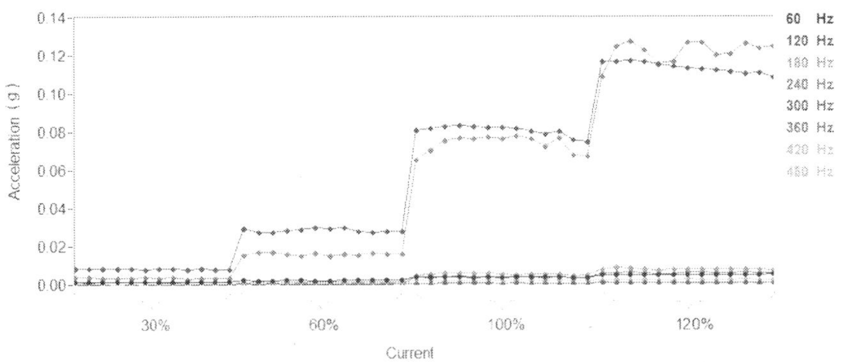

Fig. 8. Vibration signals when excited with current at sensor 2 in all frequencies.

Figures 7 to 10 show some examples of the experiments corresponding to cold transformer excited first with current and no voltage, and then excited with voltage and no current, i.e., windings excited or core excited. The vertical axis represents the magnitude of the vibration measured in terms of acceleration and expressed in g, the gravity. The horizontal

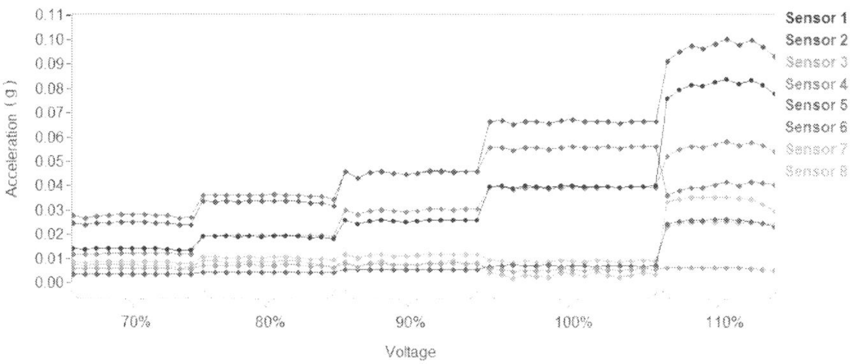

Fig. 9. Vibration signals when excited with voltage at 120 Hz. in all sensors.

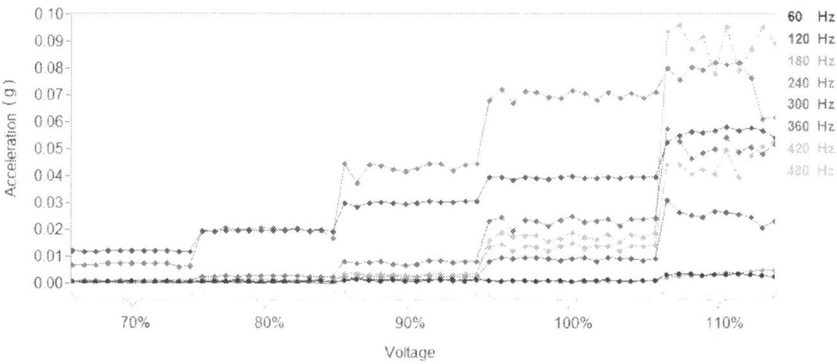

Fig. 10. Vibration signals when excited with voltage at sensor 2 in all frequencies.

axis represents each one of the ten (or twelve) repetitions of each experiment with the same operational condition.

Figure 7 shows the vibration signals when excited with current at 120 Hertz in all sensors. Notice that the steps shown in the figure correspond to excitations of 30% of the nominal current (lower amplitudes) and then 60%, 100% and 120%. Figure 8 shows the vibration signals captured at sensor 2 in all the frequencies of the same experiment. Notice that the amplitude of the vibration increases when current increases. Notice also that the frequencies of 120 and 240 Hertz are the only representatives of the vibrations compared to other multiples of 60 Hertz.

Figures 9 and 10 show the experiments with voltage and no current. Figure 9 shows the vibration signals at 120 Hetrz in all sensors, and Fig. 10 shows the vibration at sensor 2 in all frequencies.

These graphs are examples of the kind of variations that we found in the vibrational pattern, under different operational conditions.

Following the transformation of the vibration signals in their frequency components, a normalization procedure is applied. Normalization in this context means that all variable values lie between 0 and 1. This is because we only need to compare the behavior between all the vibration signals. The normalization is obtained dividing all the vibration signals by the highest measure of each sensor. Figure 11 shows an example of normalized signals. Notice that all signals detected at all sensors behave similar even if their amplitude are different as was shown in Fig. 7.

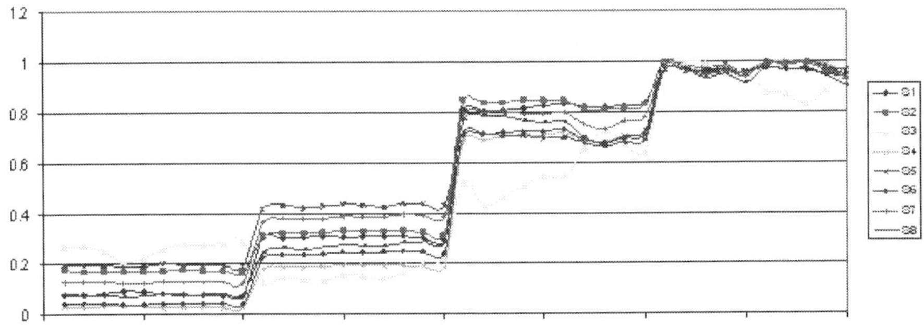

Fig. 11. Comparison between the behavior of all the signals when normalized.

Finally, a discretization is required since the probabilistic model utilizes Bayesian networks with discrete signals. Discretization is the division of the complete range of values in a fixed number of intervals. In our experiments, the vibration signals were discretized in 20 intervals or *states* $S_0, S_1, \ldots, S_{19}$. Since normalized, the states consists in 5% of the normalized signals, i.e., $0 - 0.05$, $0.05 - 0.1$ and so on.
Table 3 resumes the variables utilized in the diagnosis and the values that they can take.

Variable	*Values*
Temperature	Cold, hot
Excitation	Voltage, Current
Nominal Voltage	70%, 80%, 90%, 100%, 110%
Nominal Current	30%, 60%, 100%, 120%
Sensors	A1, A2, ..., A8
Frequencies	60 Hz., 120 Hz., 180 Hz., ..., 900 Hz., 960 Hz.

Table 3. Variables utilized in the diagnosis.

The next section utilized these variables to build the probabilistic models.

4.2 Model of correct transformers

In the first stage of this project, the variables available for constructing the model are sensors, frequencies, temperature and excitation of the transformer (voltage or current). Following the experts' advice, we consider two possible set of models. The first is a model relating operational conditions and frequencies. One model for each sensor. The second possible set of models relates operational conditions and sensors. One model for each frequency. We decided to try a set of models that relates conditions and sensors, i.e, operational conditions and vibrations detected in certain parts of the transformer. Figure 12 shows one instance of the resulting model.

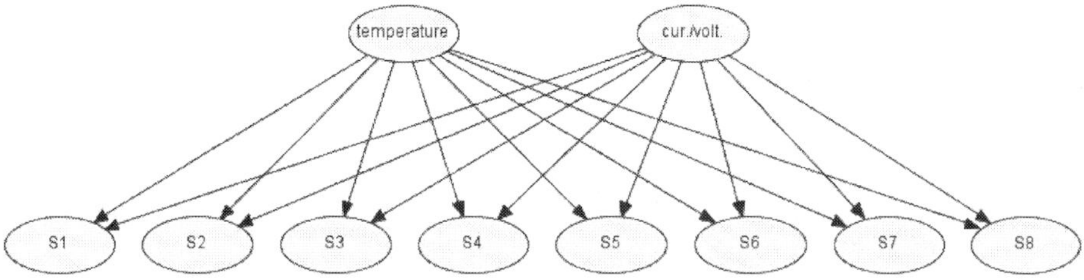

Fig. 12. Model that relates operational conditions with the amplitude measured by each sensor.

Actually, the complete model is formed by two BNs like the one shown in Fig. 12. One corresponding to the 120 Hz component and the second corresponding to 240 Hz. Once defined the structure, the EM (Estimation-Maximization) algorithm (Lauritzen, 1995) is utilized to obtain the conditional probability tables. We used 10 experiments of each type as indicated in Table 2 and applied in 5 transformers. The structure and the parameter learned, complete the models for the diagnosis. Next section describes the diagnosis procedure in the factory floor.

4.3 Diagnosis procedure in FAT

Utilizing the models described above, the algorithm 1 is applied to identify abnormal vibrations in the sensors given certain operational conditions:

Algorithm 1 Detection of abnormal vibrations.

Require: Operational conditions of temperature and excitation.
  assign a value (instantiate) to the *t*emperature and *e*xcitation nodes
  for all sensors (frequencies) in the network **do**
    propagate probabilities and obtain a posterior probability of all sensors (frequencies) nodes
    compare the real value measure and the estimated value
    evaluate if there is an error in the sensor (frequency)
  end for

As an example, Table 4 shows the measures that have been obtained and normalized in the sensors of a cold transformer excited with 100% of nominal current.

Sensor 1	Sensor 2	Sensor 3	Sensor 4	Sensor 5	Sensor 6	Sensor 7	Sensor 8
0.3284	0.3710	0.0895	0.4161	0.0811	0.7084	0.6531	0.2333

Table 4. Example of vibration measured in the sensors.

According to the algorithm 1, one sensor vibration is estimated using the rest of the sensor signals and the operational conditions. The probabilistic propagation in the BN produces a posterior probability distribution of the estimated sensor value. The problem is to map the observed value and the estimated value to a binary value: $\{correct, faulty\}$. For example, Fig. 13 left shows an example of a posterior probability distribution, and Fig. 13 right shows a wider distribution. In both cases, the observed value of the estimated sensor is shown by an arrow. Intuitively, the first case can be mapped as correct while the second can be taken as erroneous.

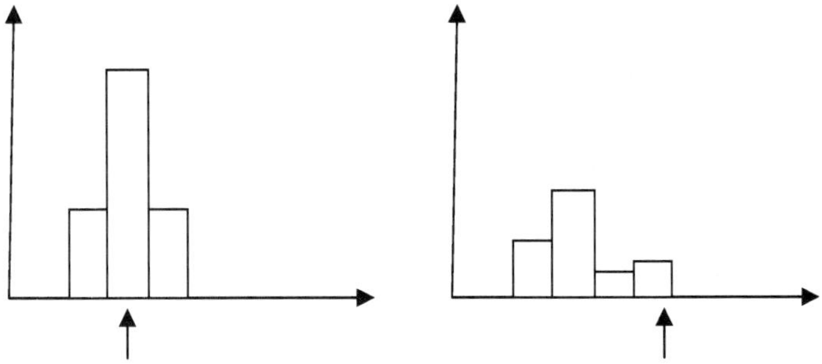

Fig. 13. Example of two posterior probabilistic distributions and the comparison with the value read.

In general, this decision can be made in a number of ways including the following.

1. Calculate the distance of the real value from the average or *mean* of the distribution, and map it to faulty if it is beyond a specified distance and to correct if it is less than a specified distance.
2. Assume that the sensor is working properly and establish a confidence level at which this hypothesis can be rejected, in which case it can be considered faulty.

The first criterion can be implemented by estimating the mean μ and standard deviation σ of the posterior probability of each sensor, i.e., the distribution that results after the propagation. Then, a vibration can be assumed to be correct if it is in the range $\mu \pm n\sigma$, where $n = 1, 2, 3$. This criterion allows working with wider distributions where the standard deviation is high and the real value is far from the mean μ value as shown in Fig. 13 right. However, this technique can have problems when the highest probability is close to one, i.e., the standard deviation is close to zero. In such situations, the real value *must* coincide with that interval.
The second criterion assumes as a *null-hypothesis* that the sensor is working properly. The probability of obtaining the observed value given this null-hypothesis is then calculated. If this value, known as the *p-value* (Cohen, 1995), is less than a specified level, then the hypothesis is rejected and the sensor considered faulty. Both criteria were evaluated experimentally. Here, it is worth mentioning that using the *p-value* with a 0.01 rejection level, works well.

4.4 Experiments for FAT

We designed a computational program that utilize the measurements obtained in the experiments described in Table 2. We run experiments and identify if there is a failure.
An experiment consists in establishing the operational conditions of excitation and temperature. Next, the system obtain the measurements of the sensors, and executes the

algorithm for detection of abnormal vibrations. Since we have 16 indicators of fault (eight sensors in two frequencies), a sensor fusion technique is required. We decided to make a weighted sum of the conclusion of each sensor. If one sensor is sensible and detects deviations easily, then a low weight is assigned. Other less sensible sensors may have higher weights. Given the sum of each frequency, we can configure ranks for declare a transformer as {correct, suspicious, faulty}.

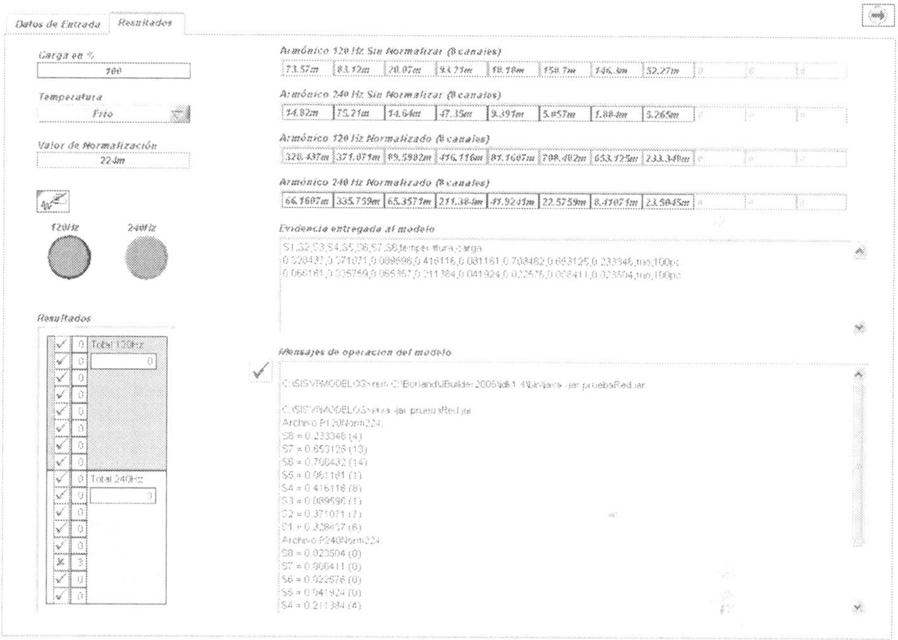

Fig. 14. User interface of the diagnosis software (in Spanish).

Figure 14 shows the results of one experiment (in Spanish). In the upper left of the window, the operational conditions are indicated. First, load *(carga en %)* with 100% of current, and cool transformer temperature *(frio)*. In the middle left of the window, there are two lights. One corresponds to a model for 120 Hz. and the other corresponds to 240 Hz. As mentioned above, we are using one model for each frequency. These lights become green if the transformer is correct, yellow if the transformer is suspicious and red if there is definite a failure. Below, in the lower left of the window, there are a little box for each sensor in the transformer. The first 8 corresponding to 120Hz and the last corresponding to 240 Hz. If the posterior probability obtained in a node (sensor) corresponds to the vibrational value currently detected, then an OK mark is described, and a NO-OK mark otherwise. Notice that the sixth sensor detected a deviation in the model of 240 Hz. In the upper right of the window, four rows of data are included. The first two correspond to the current vibration amplitude measured in the 8 sensors in the transformer. The next two rows correspond to the normalized information. They are actually the inputs to the BNs. The lower right part of the window displays other prototype information.

Several transformers have been tested in factory and some faults have been detected. The next section describes the changes made to the model in order to run SAT tests.

4.5 Preliminary experiments for SAT

Experiments in site have certain differences with FAT experiments. The main difference is that transformers always operate at their fixed nominal voltage but variable current. The current value corresponds to the demanded power by the consumers.
In order to utilize the information acquired in FAT experiments, one assumption was necessary: vibration corresponds to the sum of vibration by current (produced at the winding) plus vibration by fixed voltage at 100% of nominal value (produced at the core). This assumption is valid at the transformer operational condition below the saturation condition. Voltage is always fixed at its nominal value (controlled by the grid), and the current is always tried to keep in normal conditions. In reality, we use all the information acquired for FAT experiments, modified with this assumption.
Additionally, we run experiments in power transformers working on site. Of course, we could not modify the working conditions and we took only data in certain loads.
Table 5 shows an example of the experiments carried out at the power transformer in Prolec GE substation. The transformer provides power to the entire plant. Columns indicate the measurement obtained by all every sensor. The first row indicates the real amplitude obtained by the sensor and normalized. Once normalized, the signals are discretized in 20 intervals. The second row indicates the interval number, from 0 to 19. Third row indicates the posterior probability obtained after the propagation in the probabilistic model. This number indicates the probability of being a normal measurement, so the fourth row decides if there is a failure (1 value) or there is no failure (0 value). This decision is based on the assumption that the posterior probability distribution is Gaussian given certain operational conditions. Thus, the real value measured is compared with $\pm\sigma$ from the media.

	S1	S2	S3	S4	S5	S6	S7	S8
Real value measured	0.159	0.121	0.184	0.178	0.083	0.016	0.729	0.141
Corresponding interval	3	2	3	3	1	0	14	2
Posterior probability	0.312	0.312	0.0	0.0	0.687	0.0	0.0	0.375
Decision	0	0	1	0	0	0	1	0

Table 5. Example of one experiment.

For example in Table 5 , sensor 2 measured a normalized value of 0.121 that corresponds to the interval number 2. Propagation indicates 31% of the value that corresponds to no failure. On the contrary, sensor 7 reads a normalized value of 0.729, corresponding to interval 14 and there is no probability of being correct. The decision is 1. Notice however, that sensor 6 has the same 0 probability but the standard deviation may be very wide and the decision marked 0.
The prototype was constructed using the *hugin* platform (Andersen et al., 1989), so the off–line automatic learning and the on–line propagation are carried out with the Java APIs of this package.
Several tests were made in this Prolec GE substation transformer and the model resulted in a correct tool for transformer diagnosis.

5. Conclusions and future work

The main contribution of this work is the construction of a probabilistic vibration model obtained with the vibration signals measured in a power transformer. Thus, if a model of correct behavior can be obtained, then early deviations of this behavior can also be achieved. Our approach utilizes Bayesian networks as the formalism for constructing and utilizing the models. We used 8 sensors situated all around the tank of the transformer. Every measure was transformed to the frequency domain and only amplitude multiples of the 60 Hz were considered. Experiments were carried out at different operational conditions to construct the models. Finally, a diagnosis program receives vibration data from a transformer, inserting it as evidence and probability propagation allows calculating the probability of proper behavior. Bayesian networks have the advantage of generate conclusions even when the evidence is incomplete. This means that even with less sensors or less frequencies, a conclusion can be obtained. Also, BNs include several algorithms that automatically adapt the models, based on vibration in the normal life of the transformer. This means we can detect the normal behavior of old transformers even if they vibrate much more that their vibration when new.

Future work is needed in the determination of additional operation conditions variables, like parameters in the construction of each transformer. We can detect the vibration transmission between different parts of the transformer and identify more clearly if the behavior is normal or not.

Final results will be available after months of tests in new and old transformers, in site and at the factory.

6. Acknowledgments

This research is partially supported by Consorcio Xignux-Conacyt and by the Prolec GE-IIE project 13261-A.

7. References

Andersen, S. K., Olesen, K. G., Jensen, F. V. & Jensen, F. (1989). Hugin: a shell for building bayesian belief universes for expert systems, *Proc. Eleventh Joint Conference on Artificial Intelligence, IJCAI*, Detroit, Michigan, U.S.A., pp. 1080–1085.

CFE (2010). Generación termoeléctrica, http://www.cfe.gob.mx/quienessomos/estadisticas/Paginas/Indicadoresdegeneracion.aspx.

Cohen, P. (1995). *Empirical methods for artificial intelligence*, MIT press, Cambridge, Mass.

Crowley, T. H. (1990). *Automated diagnosis of large power transformers using adaptive model-based monitoring*, Master of science in electrical engineering, Massachusetts Institute of Technology, MIT, Boston, Mass., U.S.A.

García, B., Burgos, J. C. & Alonso, A. M. (2006a). Transformer tank vibration modeling as a method of detecting winding deformations - part i: Theoretical foundation, *IEEE Transactions on Power Delivery* 21(1): 157–163.

García, B., Burgos, J. C. & Alonso, A. M. (2006b). Transformer tank vibration modeling as a method of detecting winding deformations - part ii: Experimental verification, *IEEE Transactions on Power Delivery* 21(1): 164–169.

Golubev, A., Romashkov, A., Tsvetkov, V., Sokolov, V., Majakov, V., Capezio, O., Rojas, B. & Rusov, V. (1999). On-line vibro-acustic alternative to the frequency response analysis and on-line partial discharge measurements on large power transformers, *Proc. TechCon Annu. Conference*, TJ/H2b, Analytical Services Inc., New Orlans, L.A., U.S.A., pp. 155–171.

Harlow, J. H. (2007). *Electric Power Transformer Engineering*, CRC Press.

Lauritzen, S. L. (1995). The em algorithm for graphical association models with missing data, *Computational Statistics & Data Analysis* 19: 191–201.

Lavalle, J. C. (1986). *Failure detection in transformers using vibrational analysis*, Master of science in electrical engineering, Massachusetts Institute of Technology, MIT, Boston, Mass., U.S.A.

McCarthy, D. J. (1987). *An adaptive model for vibrational monitoring of power transformers*, Master of science in electrical engineering, Massachusetts Institute of Technology, MIT, Boston, Mass., U.S.A.

Neapolitan, R. (2004). *Learning Bayesian Networks*, Prentice Hall, New Jersey.

Pearl, J. (1988). *Probabilistic reasoning in intelligent systems: networks of plausible inference*, Morgan Kaufmann, San Francisco, CA.

7

Measurement of Satellite Solar Array Panel Vibrations Caused by Thermal Snap and Gas Jet Thruster Firing

Mitsushige Oda[1], Yusuke Hagiwara[2], Satoshi Suzuki[3], Toshiyuki Nakamura[1], Noriyasu Inaba[1], Hirotaka Sawada[1], Masahiro Yoshii[1] and Naoki Goto[1]

[1]Japan Aerospace Exploration Agency (JAXA)
[2]Tokyo Institute of Technology
[3]AES Co., Ltd
Japan

1. Introduction

Many space satellites have large solar array paddles (Fig. 1) for power generation and large antennas for observation and communication. These large space structures are folded during transport into space by launch vehicles, and deployed after arriving in space. The paddles and antennas must be lightweight because of the payload weight limit of the launch vehicle and are therefore very flexible, with little damping ability. This results in vibrations, which cause serious problems. In particular, there have been increasing demands for enhanced resolution of Earth observations from low Earth orbiting satellites in recent years. Accordingly, the requirements for satellite attitude stability are also increasing. Conversely, it is also known that the attitude stability of low Earth orbiting satellites is disturbed when the satellites go into and leave an eclipse. When the thermal environment around a flexible structure on orbit such as a solar array paddle changes to cold or hot, the flexible structure

Fig. 1. A solar array paddle in orbit © JAXA

produces its own deformation or vibration. These occur most often during rapid temperature changes called thermal snap or thermally-induced vibration, which has been known to cause attitude disturbance in Low Earth Orbit (LEO) satellites.

Thermal snap vibration occurring on a flexible solar array panel is very slow, and measuring motion on a solar array panel caused by thermal snap with sensors, such as an accelerometer, is very difficult. The behaviour of a space structure affected by thermal snap has never been observed directly in space until now. In this chapter, our vibration measurement method, along with images taken in space and the image processing conducted with the images taken on the ground, is explained, and some measurement results are shown.

2. Thermal snap

Thermal snap is a unique phenomenon affecting flexible space structures and one of the main factors causing disturbances in satellites. When a satellite enters an eclipse, the whole of the solar array paddle or solar array panels are subject to rapid cooling, while the temperature of the panel on the sun side is high beforehand. When the satellite leaves the eclipse, the sun side of the solar array panels heats up rapidly, while the temperature difference between both sides of the solar array panel is small when the satellite is in the eclipse.

These rapid changes of temperature difference between both sides of the solar array panels result in thermal expansion and shrinking of the solar array panels and cause the solar array panels to bend. This bending then, in turn, causes the solar array panels to vibrate and results in degradation of the satellite attitude stability.

Many past satellites and space structures observed thermal snap from their attitude data. The Hubble Space Telescope (HST) shown in Fig. 2 was launched by the space shuttle *Discovery* on April 25, 1990, to conduct higher-accuracy astronomical observations than ground-based equipment (Foster et al., 1995). A pointing control system of HST was designed to hold an image stable at the HST focal plane to 0.007 arcsec (rms) for the duration of an observation. However, following successful deployment of a pair of solar array paddles, gyro data revealed significant attitude disturbances; observed when HST entered or left the shadow of the earth. Based on investigation of HST's telemetry data and certain analyses, it was concluded that the disturbance of HST was caused by the thermally-induced deformation of the HST solar array paddle. The HST solar array paddle has a flexible solar array blanket and deploys a boom named the two-element Storable Tubular Extendible Member (Bi-STEM). When HST remains in orbit on the day side, its deploying boom heats up and a thermal gradient appears. The surface directed toward the Sun is hotter than the opposite one and the thermal gradient also causes the solar array paddle to bend. Conversely, when HST is in orbit on the night side, the whole of the boom cools down and the thermal gradient disappears, while the bending of the solar array paddle is also absent. This bending motion occurs rapidly during sunshine/eclipse transition and disrupts HST's pointing system. In December 1993, the space shuttle *Endeavour* was launched and the solar array paddles were replaced with new units to counter the problem.

The Upper Atmosphere Research Satellite (UARS), launched in September 1991, and the Advanced Land Observing Satellite (ALOS), launched in January 2006, in addition to HST, also observed an attitude disturbance of the main body based on gyro telemetry data when they traversed boundaries between the orbital day and night sides (Iwata et al., 2006; Johnston & Thornton, 2000). However, these phenomena are rarely observed or measured

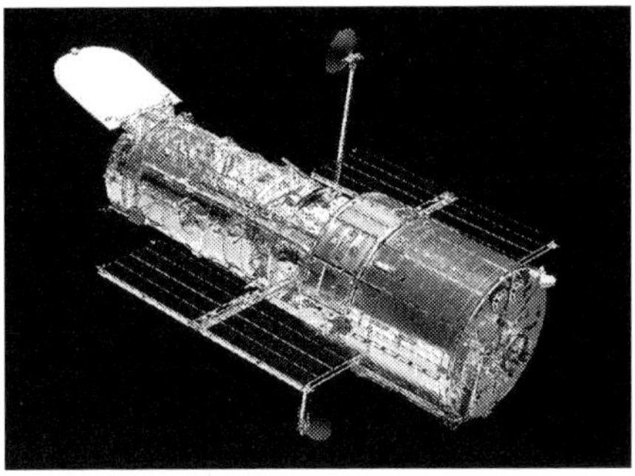

Fig. 2. Hubble Space Telescope

directly since the solar array panel motions caused by thermal snap are very slow and difficult to measure using sensors such as accelerometers. Additionally, the measurement system has to be compact because of the strict payload limit, meaning there have been hardly any direct measurements of thermal snap in orbit to date.

3. Measurement system using an onboard monitoring camera

This section introduces the equipment and configuration of our measurement system. To measure vibration on flexible space structures, we proposed a method using a single monitoring camera mounted on a satellite, the latter of which has several monitoring cameras to monitor its own states. Our measurement system uses one of them, which monitors solar array paddles.

3.1 GOSAT

Our measurement system is installed on JAXA's (Japan Aerospace Exploration Agency) Earth observation satellite named GOSAT (Greenhouse gases Observing Satellite), and measures the deformation and vibration of its solar array paddle. GOSAT was launched on January 2009. Its major parameters are shown in Table 1, and an overall view in Fig. 3.

Size	Main body	3.7 m (H) × 1.8 m (W) × 2.0 m (D)
	Wingspan	13.7 m
Mass		1750 kg
Power		3.8 kW
Lifespan		5 years
Orbit		Sun Synchronous Orbit
	Local time	13:00 ± 0:15
	Attitude	666 km
	Inclination	98 deg
	Re-visit	3 days
Launch	Vehicle	H-IIA
	Schedule	Jan. 2009

Table 1. GOSAT Summary

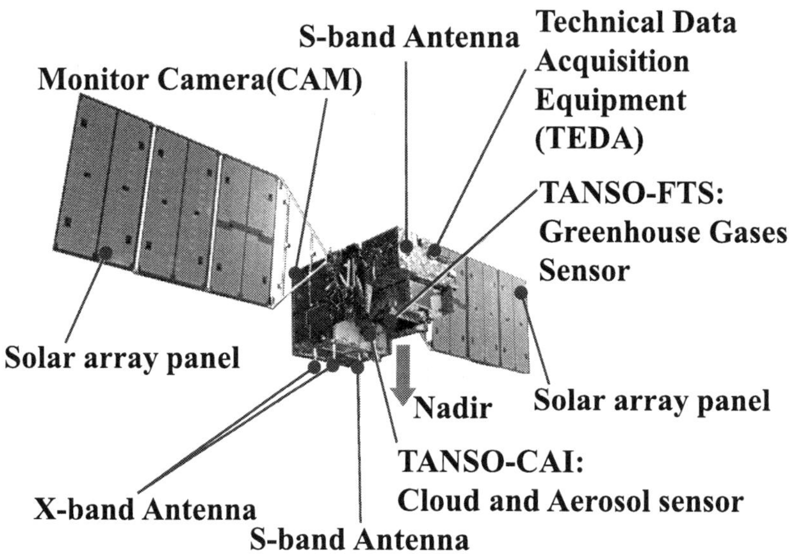

Fig. 3. Overall view of GOSAT

GOSAT has eight CMOS cameras to monitor the deployment state of solar array paddles, their post-deployment behaviour, the existence of contamination in the opening of fairing, separation of the satellite from a launch vehicle, and so on. The effective sensor resolution of the cameras is 1.3 million pixels. The monitoring camera which we used for monitoring the whole of the solar array paddle thus has a short focal length and a wide view angle while the monitoring camera we used for measurement is shown in Fig. 4 and has an LED lighting system.

Fig. 4. Monitoring camera mounted on GOSAT © Meisei Electric Co., Ltd.

3.2 Assumptions to measure the vibration

The solar array panel's vibration is measured using one CMOS camera mounted on the satellite main body. Since the images taken by a single CMOS camera do not include 3-D information, some assumptions are needed to identify the vibration of the solar array panels. The major assumptions are as follows:

3.2.1 Solar array panel

The GOSAT has a pair of solar array paddles, unlike most remote sensing satellites in low Earth orbits, which have only one. This is to improve robustness against failures in the solar array paddle, which resulted in the total loss of the ADEOS and ADEOS-II, which are JAXA's former remote sensing satellites. Each paddle consists of three semi-rigid panels and a yoke. These solar panels were folded down to the satellite main body during the rocket launch and released in orbit. During this measurement, we consider the solar paddle, which comprises three panels, to be actually a single uniform panel. Fig. 5 shows the size of the solar array paddle and its visual markers, namely about 2800 × 6000 mm.

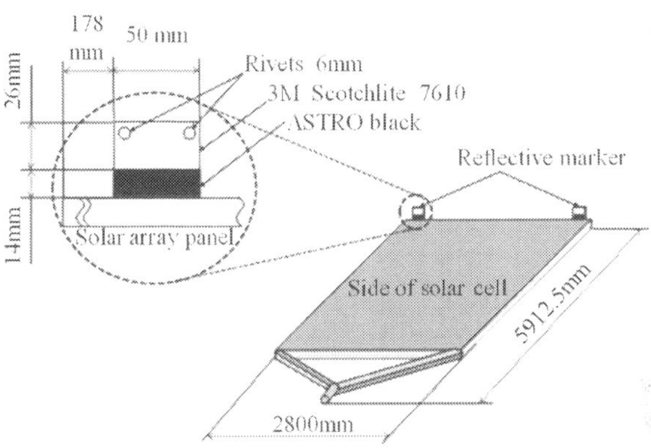

Fig. 5. Size of the solar array paddle and its visual marker

3.2.2 Visual markers

The vibration of the solar array paddle caused by thermal snap occurs during poor lighting conditions for the monitoring camera (eclipse). Additionally, the lighting condition changes dramatically while the satellite goes from sunshine into the eclipse. Visual markers and an LED lighting system are used to help Interference the motion of the solar array paddle. The visual marker is made of reflective tape and is 50 × 26 mm in size. Unfortunately, these markers are attached only to the front surface (solar cell side) of the solar paddle, to avoid mechanical interface with the solar cell panels when they are folded down to the satellite main body. Displacement of the solar array paddle can be obtained by measuring the positions of the Visual markers.

3.2.3 Vibration modes

We assumed that the GOSAT's solar array paddle has three major vibration modes as shown in Fig. 6, namely out-of-plane vibrations, twists, and in-plane vibrations. Higher order vibrations are not considered.

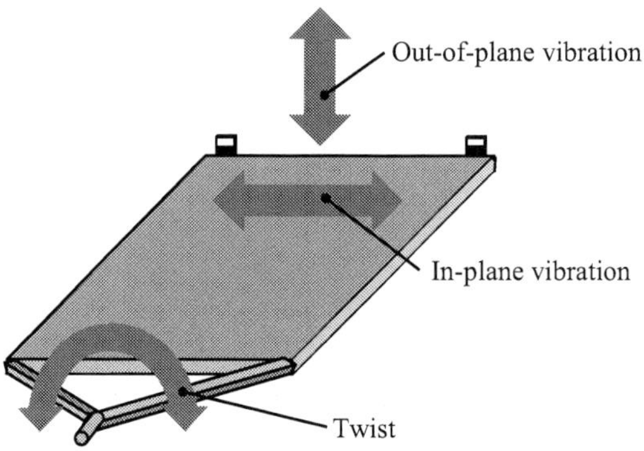

Fig. 6. Vibration modes of the solar array panel

3.3 Ground-based test model for the algorithm development

To develop the measurement system and algorithm, a ground-based test model comprising a scale model (1/3 scale) of the solar cell paddle, a compatible CMOS camera and an LED lighting system are developed. An artist's image of the test bed is shown in Fig. 7, and that of the solar array paddle taken by the ground-based test model is shown in Fig. 8.

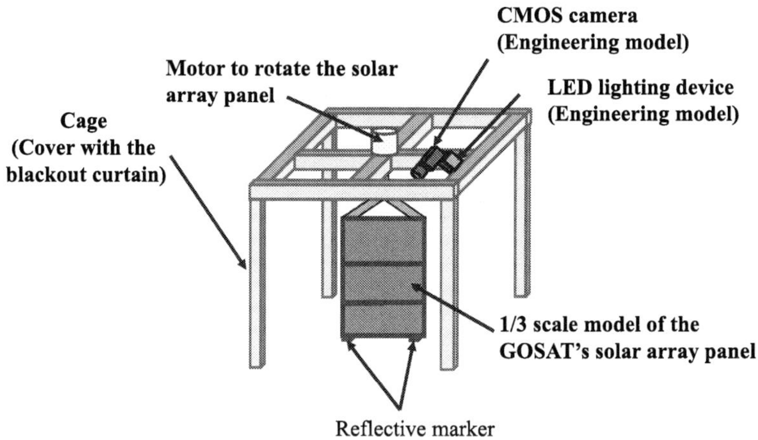

Fig. 7. A ground-based test model to develop the measurement system

4. Development of a high-accuracy image processing algorithm

Images taken by the mounted camera are down-linked to the ground and subject to processing to measure the position of the reflective markers. The algorithm for the image processing is shown in this section. One problem when measuring thermal snap is that the lighting condition changes dramatically during the measurement. To resolve this, the image processing algorithm is adjusted to use the same algorithm for both sunshine and eclipse. A ground-based test model was used to develop our measurement system.

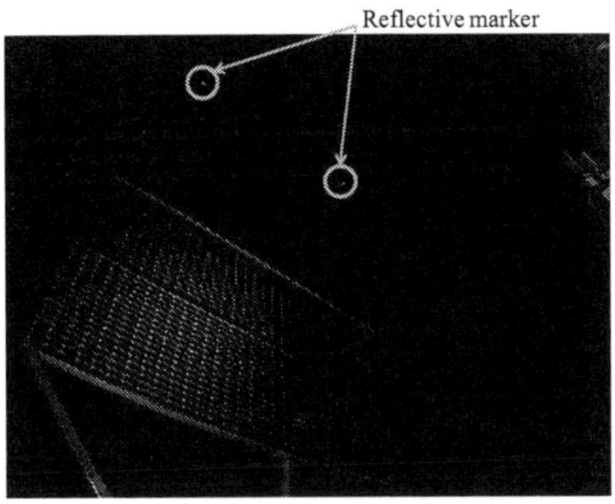

Fig. 8. Image of the solar array paddle taken by the test bed

4.1 Calibration of camera parameters

Since the CMOS camera optical lens system is usually simple, the images taken always include distortion, which must be corrected. Image data processing, such as searching the visual marker, is based on MVTech's HALCON system.

The camera parameters are necessary to obtain the marker position by image processing and include internal and external camera parameters. The internal camera parameter comprises eight items of Focus, Kappa, Sx, Sy, Cx, Cy, Image width and Image height. Focus is the focal length of the lens. Kappa is the distortion coefficient of the lens. Sx and Sy are the distances between the cells. Cx and Cy are coordinates at the distortion centre. The external camera parameter shows the relation between the measurement plane and the camera (position and orientation). This camera parameter can be easily obtained by using the camera calibration program installed in HALCON.

Fig. 9 shows the standard calibration table of HALCON, which is used for its calibration. Since the black spot of the standard calibration table interval is already known, the camera parameter can be obtained by taking a photograph of the standard calibration plate.

Fig. 10 shows the image of the calibration. A standard calibration table is taken of a photograph 20 times while changing the relation of the camera (position and orientation). Afterwards, the standard calibration table is set up in the measurement plane, and a photograph is taken. The internal and external camera parameters were calculated using these 21 images.

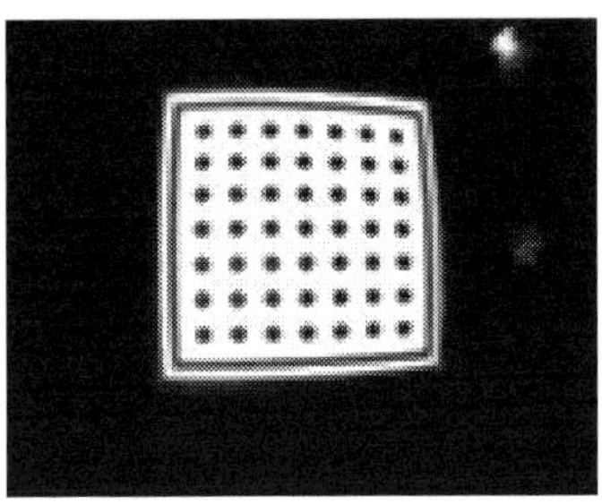

Fig. 9. Calibration table

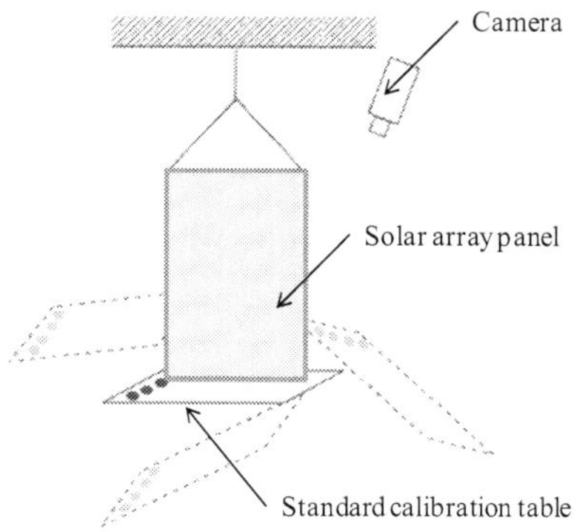

Fig. 10. Calibration method using a standard calibration table

4.2 Development of the image processing algorithm

Fig. 11 is the flow chart used to obtain the position of the marker from the image. Initially, the image is converted into gray scale, whereupon the marker candidate area is searched for, using the entire image. The edge extraction processing is then performed in the marker candidate area, with the marker and noise distinguished by the length, size, and circularity of the extracted edge. If the number of edges that remove the noise is two, the edges are considered markers. Subsequently, the centre of the marker is obtained in the image coordinate system [Row, Column] (pixel) and converted into a world coordinate system [X, Y] (mm) by the camera parameter. The distance between the two markers is calculated, and the correctness of the value is determined. When the image processing is not the first frame, the distance the marker has moved from the previous frame is calculated. If this numerical result is correct, the image processing is considered a success. When this happens, the surround of the marker position of the present frame is assumed to be a marker candidate area of the following frame.

Fig. 12 is the image processing result. The marker to the left of the screen is called No. 0 and the other marker is called No. 1. This figure shows successful marker extraction through image processing, with Fig. 13 a transition of the distance between the two markers. The accuracy of the image processing can be shown by the size of the change of the distance between the two markers, hence the standard deviation of the distance between markers was assumed to represent the image processing accuracy. In the ground experiment, this value was about 0.99 mm, which corresponds to about 1/5 of the resolution. This result shows that image processing, the accuracy of which exceeds the image resolution, is realized on the ground.

4.3 Adjusting the algorithm for edge extraction

Edge extraction is applied to find markers in our algorithm. One problem when measuring thermal snap, however, is that the lighting condition changes dramatically during the measurement. One kind of algorithm for finding markers should be used under every lighting condition. To resolve this problem, the image processing algorithm is adjusted to use the same algorithm in both sunshine and eclipse. Subsequently, the algorithm is upgraded to enable markers not only when the satellite is in the umbra but also when it is in the sunshine and penumbra, based on the algorithm for the inside of the eclipse. The following are the contents of the upgrade of the algorithm.

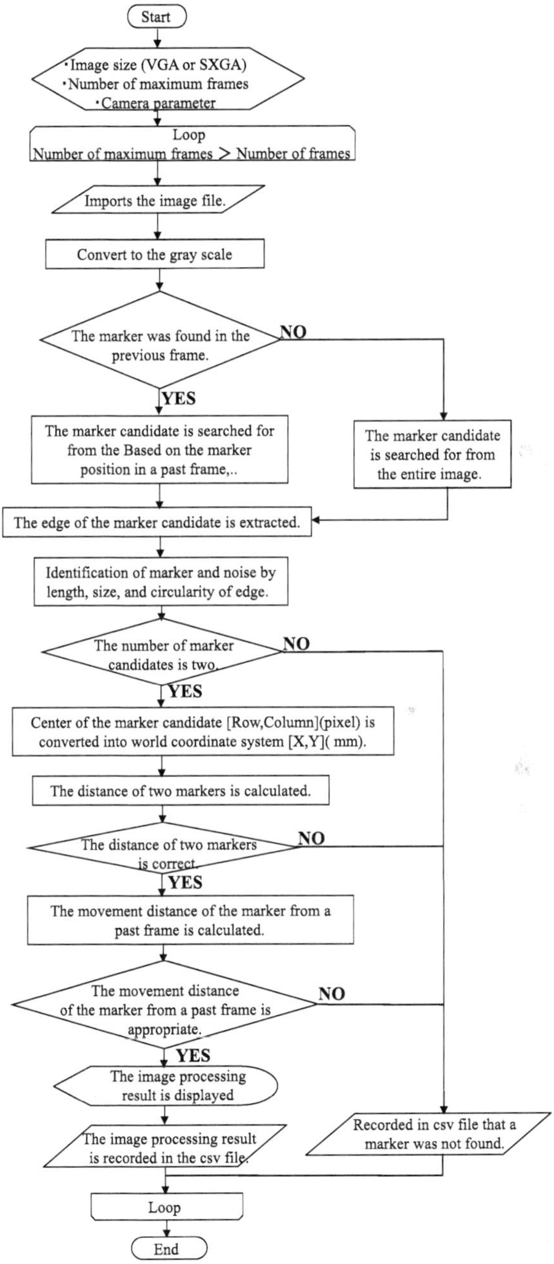

Fig. 11. Algorithm to find the marker in the images

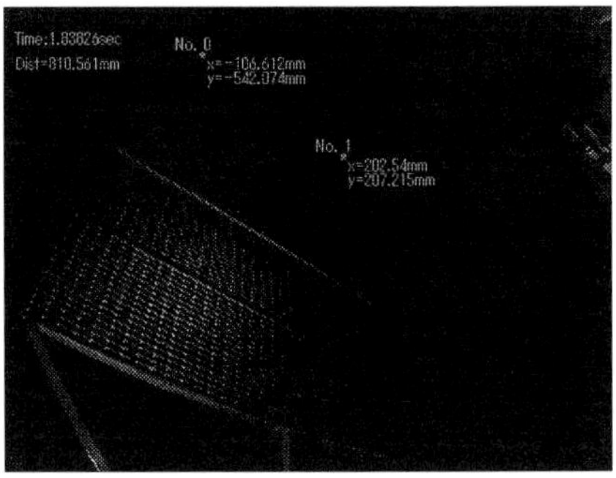

Fig. 12. The image processing result using the ground-based test model

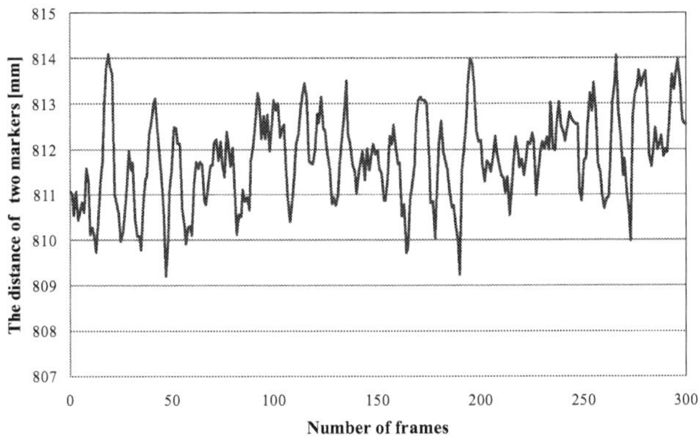

Fig. 13. Transition of the distance between two markers on the ground

4.3.1 Detecting the marker candidate area

Many lighting points have equivalent size and shape when compared to actual markers in the sunshine and penumbra. Therefore, when the analysis starts, the marker candidate area is specified manually in the new algorithm. Subsequently, if the image processing is successful, the marker candidate area in the next frame is specified at near the current marker candidate area.

4.3.2 Determining the threshold for finding markers

The thresholds for finding markers change depending on the luminance of the marker candidate area in the upgraded algorithm, since this changes around the time when the satellite goes into eclipse.

4.3.3 Parameter for distinguishing the marker and noise

The conditions for distinguishing the marker and noise, for example, the size and circularity of the extracted edges, are relaxed. This is because a true marker is often mistakenly distinguished as noise if the thresholds for finding markers are changed.

4.3.4 Specifying the marker candidate area in the next frame

The marker candidate area in the next frame is distinguished as 20 pixels of the area which surrounds the marker in the current frame. The bright solar panel surface is located near the markers when GOSAT is in sunshine. Therefore, if the marker candidate area is too large, a light point on the solar panel may be distinguished as a true marker by mistake. The marker candidate area in the next frame is set up as 13 pixels in the upgraded algorithm.

4.4 Examples of image processing results in orbit

Fig. 14 shows enlarged views of the extracted markers, the edges of which are shown here as yellow lines. The paddle and markers are effectively distinguished in all cases of sunshine, penumbra, and umbra.

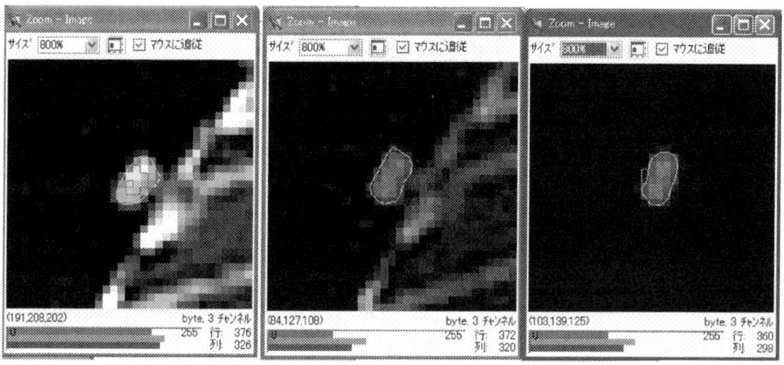

Fig. 14. Results of extracted edges (Left: Sunshine, Centre: Penumbra, Right: Umbra)

5. Measurement and analysis result of vibration using jet thrusters

When a satellite changes its orbit to increase altitude, the installed gas jet thrusters are used, and the solar array panel is subject to deformation or vibration. GOSAT has 20 newton (N) jet thrusters, and a vibration measurement was conducted when they were used. This measurement was conducted to check our measurement system before measuring thermal snap.

5.1 Measurement condition

The measurement of the solar array paddle vibration caused by 20-N thrusters was conducted during the orbital night, for a duration of about 600 seconds. Exposure of the monitoring camera was set to auto, and the resolution of the taken images was set to SXGA (1280 × 1024 pixels).

5.2 Calibration of the external camera parameters

Because the standard calibration table can be set up in the measurement plane, the external camera parameters can be easily obtained in the ground experiment, but not by the method in space. Therefore, the external camera parameters were obtained by using the image of which the GOSAT satellite had taken a photograph while in orbit. The external camera parameter is calculated from the installation position of the camera and the initial marker and is assumed to be temporary in nature. This temporary external camera parameter is corrected by the image of which the GOSAT satellite took a photograph while in orbit. First, the marker position is obtained in the image coordinate system [Row, Colum] (pixel). Subsequently, two restraint conditions are imposed on the obtained marker position [Row, Column] (pixel). One is that the distance between the markers be constant. Another is that all markers exist in the X-Y plane in the world coordinate system. The external camera parameter is corrected on the restraint condition.

5.3 Evaluation of measurement accuracy based on the distance between two markers

The internal camera parameter was obtained by a prelaunch ground experiment, while the external camera parameter was obtained by the method shown in 5.2. The algorithm shown in 4.2 was used with these parameters for the image processing.

Fig. 15 shows the transition of the distance between markers on the orbit. The average distance between markers was 2393.26 mm, and the standard deviation was 1.99 mm. The design value of the distance between markers of the GOSAT satellite is 2394 ± 2mm. The distance between markers as obtained from the image processing is within this range. The measurement plane is at a position about 5.57 m from the camera, with a resolution of about 7.25 mm/pixel. Therefore, when the standard deviation is 1.99 mm, the image processing accuracy is about 1/3.6 of the resolution.

This accuracy is about 1.5 times compared with the ground experiment result, and has decreased, seemingly due to the darkness of the image. The image darkens when the exposure is set to auto. In the ground experiment, the exposure was set to manual, and the image was processed on the condition that the marker could be subject to clear visual checks. Therefore, the image processing accuracy might improve if the exposure is appropriately set.

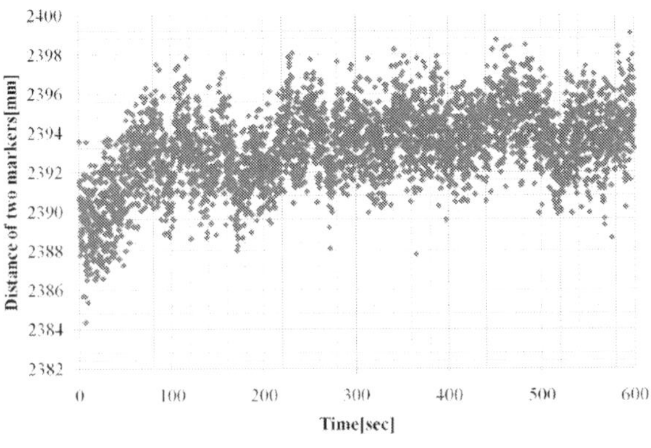

Fig. 15. The transition of the distance between two markers during the 20-N maneuver

5.4 Measurement result of in-plane and out-plane deformations

To evaluate the structural feature of the GOSAT's solar array paddle, vibration analyses are conducted using the result of the image processing conducted when the 20-N thruster was used. Three patterns of the solar array paddle's vibration modes, namely out-of-plane, in-plane, and twist, are considered. The transition of the marker position is written with the world coordinate system in Fig. 16, meaning the coordinate transformation from the world coordinate system to the local coordinate system of the solar array paddle must be conducted and the transition of the marker must be written with the solar array paddle's local coordinate system to measure the in-plane and out-of-plane vibration which occur on the solar array paddle.

Fig. 17 shows the out-of-plane and in-plane vibration of the marker No. 0. When the 20-N thruster is used, quasi-static deformation is induced while the in-plane and out-of-plane vibration occur. After the 20-N thruster, while the quasi-static deformation reverts, the vibrations continue.

The twist mode of the solar array paddle vibration can be observed based on the transition of the rotation angle of two markers. However, no deformation and vibration are observed from the transition of the rotation angle during the 20-N maneuver.

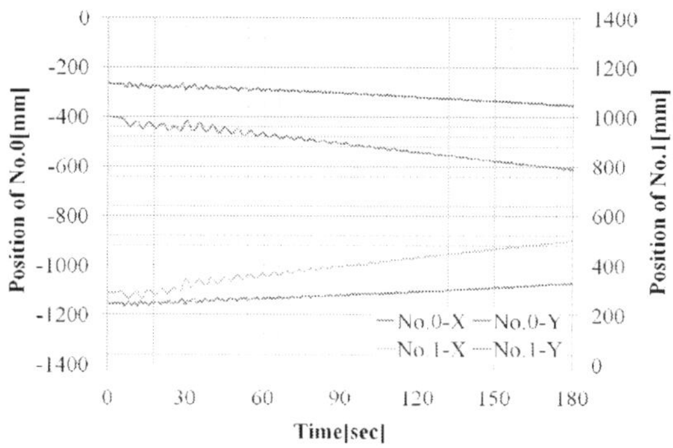

Fig. 16. The transition of the marker position as shown by the world coordinate frame

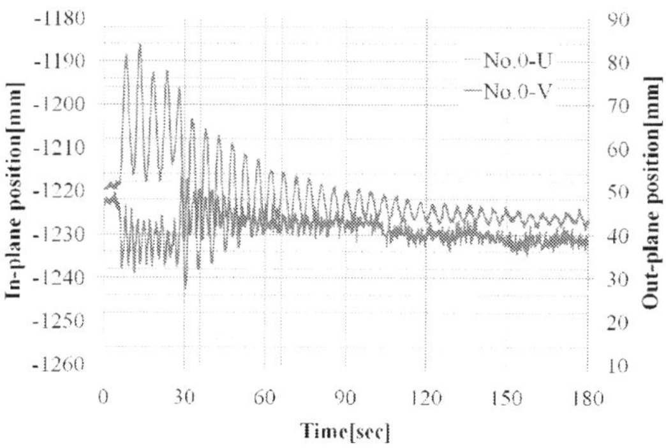

Fig. 17. Measurement result of marker No. 0

5.5 Vibration analysis

Figs. 18 and 19 shows the result of the fast Fourier transform analysis toward the solar array paddle's out-of-plane and in-plane vibration of the maker No. 0's position following the 20-

N thrust. They show that the out-of-plane vibration frequency is 0.215 Hz and the in-plane one is 0.459 Hz. Besides, both markers' in-plane oscillations are in the same phase, meaning no vibration mode, e.g. bending of the solar array panel in the direction of the panel width, occurs.

Based on the results of the fast Fourier transform analyses, 2 patterns of vibration modes can be estimated. Fig. 20 shows the estimated 2 vibration modes of the solar array paddle. The first vibration mode is the first order of the out-of-plane vibration, which is a natural frequency of 0.215 Hz. The second vibration mode is a width direction, which oscillates at 0.459 Hz.

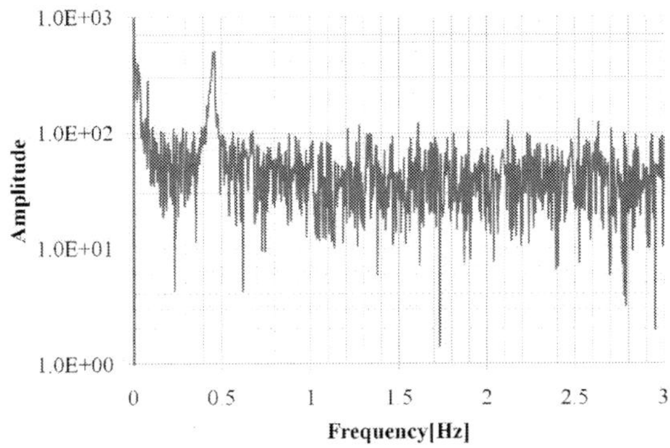

Fig. 18. Result of the FFT analysis (After maneuver, In-plane)

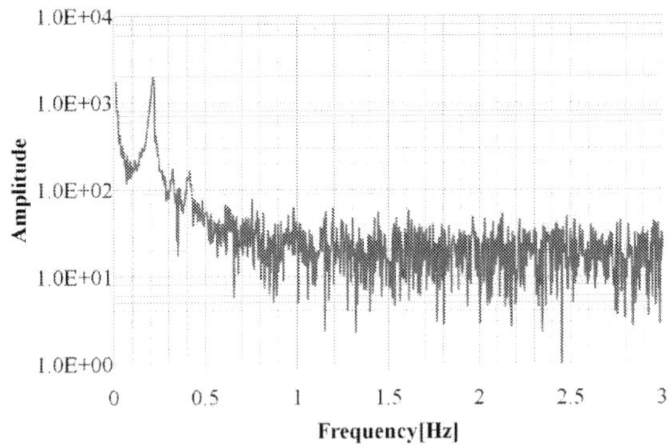

Fig. 19. Result of the FFT analysis (After maneuver, Out-plane)

5.6 Identification of the damping constant

From the out-plane deformation shown in Fig. 17, the damping constant is identified. The twelve peaks after finishing the maneuver are used for the identification. The result of the identification is 0.021, while the damping constant is so small that the natural response frequency of the out-of-plane vibration is very nearly equal to the vibration frequency.

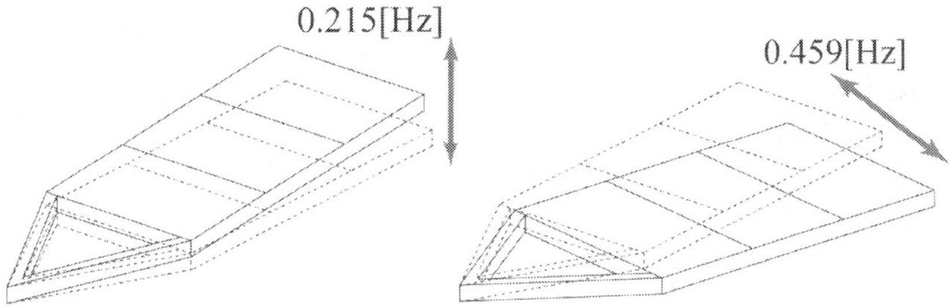

Fig. 20. Estimated vibration mode (Left: 1st. Mode, Right: 2nd. Mode)

6. Measurement result of the thermal snap

The thermally-induced deformation of the solar array paddle is measured when GOSAT goes from the sun side into the shadow of the Earth. To take good images for processing, the appropriate exposure must be set on the monitoring camera and several measurements are conducted as its exposure changes. The images taken at each measurement are subsequently processed to determine the position of the markers based on the adjusted algorithm. The distances between the two markers are evaluated and it is shown that the two markers get close during each measurement, allowing the out-of-plane deformation results of the solar array paddle to be obtained.

6.1 Measurement conditions

Fig. 21 shows the exposure and shooting time, with the LED always on, regardless of the lighting condition and the image resolution SXGA. These conditions are assigned uniform values for all measurements. Images used for measurements are taken when the satellite goes into an eclipse from the sun side. The time is recorded from the point at which the satellite enters the eclipse. The initial 10 seconds is defined as the penumbra, within which the optical environment changes momentarily. The sunshine comes before the penumbra, and the umbra starts 10 seconds after the origin of the latter. The optical environments differ dramatically between the sunshine and eclipse, making it impossible to apply uniform exposure throughout the measurement. If this is done, the brightness of the image taken in the sunshine is saturated, or an image showing nothing is produced when the satellite enters the eclipse. The exposure applied should be varied as appropriate depending on the optical environment. Therefore, several times of measurements are conducted with several exposures and several shooting times. Case 1 shown in Figure 9 is intended to take good images in the umbra, with exposure fixed to 1/16, and shooting need not be suspended to change the exposure. Cases 2, 3, and 4 are conducted to take good images in both sunshine and umbra. The optical environment in sunshine is very light, so exposure must be short enough. Exposure is initially set to 1/512, 1/1024, and 1/2048 at first in cases 2, 3, and 4 respectively. After going into the eclipse, the exposures are changed to 1/16 to take good images in the umbra in each case.

In cases 5 to 9, measurements start 2 minutes after entering the eclipse, and finish 2 minutes later in terms of elapsed time. Case 5 is conducted as a reference for the other cases, with exposure of 1/16. The exposure is set up for the range 1/32 to 1/256, and fixed in each case.

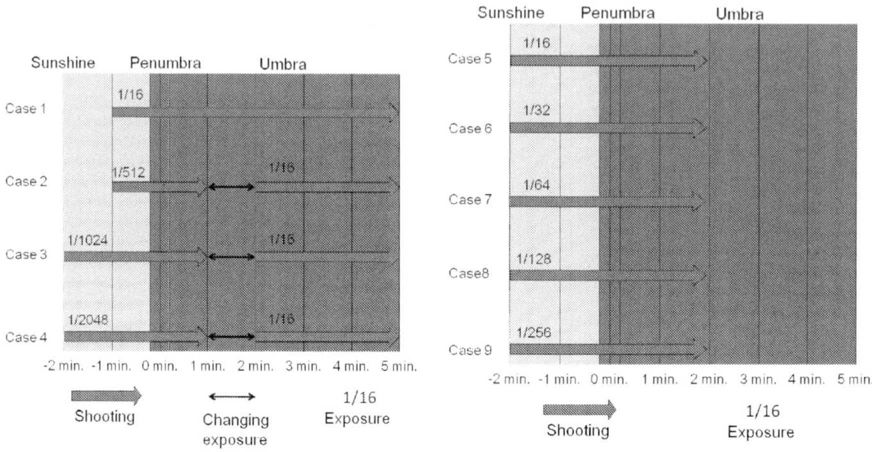

Fig. 21. Measurement condition (Cases 1 ~ 9)

6.2 Result of the thermal snap measurement

Fig. 22 shows examples of images taken in case 1. Fig. 23 is the result of cases 1, 2, 3, and 4, which shows the out-plane displacement of marker No. 0, with the upgraded algorithm for finding markers used to conduct image processing. The orange-colored line shown in Fig. 23, which shows the result of case 1, starts from around the time of origin. Therefore, the image processing succeeds, not only when the satellite is in umbra but also at the end of the penumbra. Rapid deformation occurs in the penumbra, the range of which is about 6 mm. Deformation of the solar array paddle in the umbra starts from about -4.5 mm of the out-plane displacement. Subsequently, the displacement continues to change slowly until about 120 seconds have elapsed from the start of the time elapsed, and then stops. Case 1 shows that rapid deformation occurs at the end of the penumbra, while slow deformation, which is considered quasi-static, occurs after the satellite has completely entered the eclipse. In cases 2, 3, and 4, measurements succeed in both the sunshine and umbra. The image processing is accurate to a sub-pixel level, namely sufficient. When the satellite is in the sunshine, the out-plane displacement of marker No. 0 ranges from about -10 to -7 mm, retaining nearly the same value in each case. Once the satellite enters the eclipse and the exposures are changed, the measurements are conducted again.

The out-plane displacements in the umbra are nearly the same as that of in case 1, but differ when values in the sunshine and umbra are compared. Therefore, the deformation of the solar array paddle is considered to occur from the point the satellite is in the sunshine to that when that is in the umbra.

From cases 1, 2, 3, and 4, displacement in the sunshine and umbra could be respectively obtained. However, displacement in the penumbra could not be measured well due to inappropriate exposure. Fig. 24 shows the results of cases 5, 6, 7, 8, and 9, as well as the out-plane displacement of marker No. 0. Image processing to find markers succeeded from the starting penumbra to the end of measurement in the umbra. The accuracies of each measurement are about plus or minus 1mm, which is sub-pixel level and sufficiently accurate, allowing deformations in the penumbra to be correctly determined. It is shown that rapid deformation resembling vibration occurs at the solar array paddle of GOSAT while the satellite traverses the penumbra.

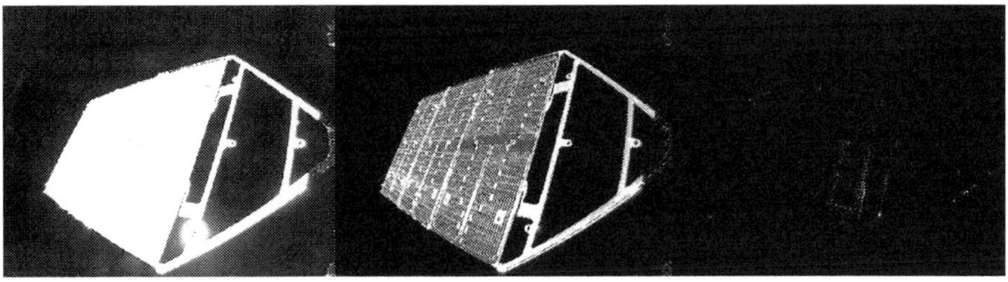

Fig. 22. Examples of images in case 1 (Left: Sunshine, Centre: Penumbra, Right: Umbra)

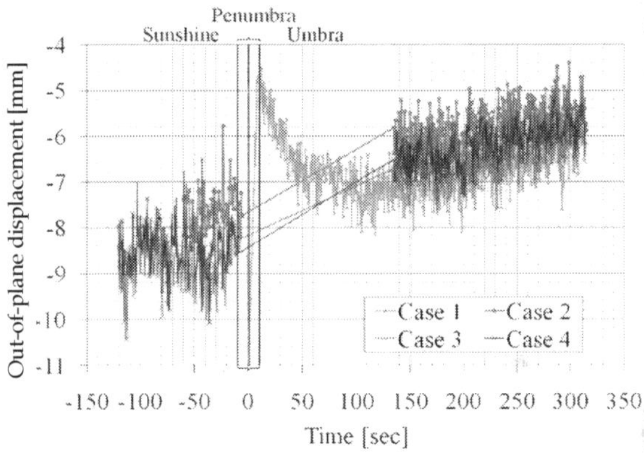

Fig. 23. Out-of-plane displacements of marker No. 0 in cases 1, 2, 3, and 4

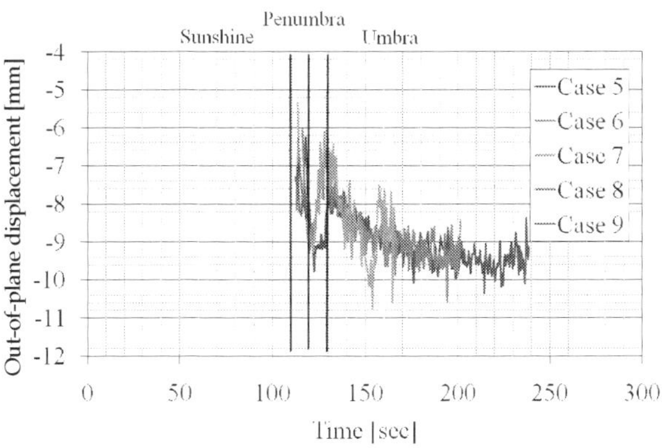

Fig. 24. Out-of-plane displacement of marker No. 0 in cases 5, 6, 7, 8, and 9

7. Conclusion

In this chapter, a measurement system using an on-board monitoring camera to observe thermal snap in orbit was explained. The thermal snap phenomenon, which causes attitude disturbance of LEO earth observation satellites, has long been observed from gyro telemetry data. However, direct observation of the thermal snap had never succeeded until now. A camera mounted on GOSAT took images when the 20-N thrusters were used and GOSAT went into the eclipse from the sunshine area, and image processing is conducted to measure the displacement of the solar array paddle of GOSAT. To obtain the result around the time of entering the eclipse, the algorithm for extracting the edges of the markers was adjusted by changing the method of detecting the marker candidate area, and the threshold for finding the markers. Based on the measurement during the 20-N maneuver, it was shown that our measurement system could realize sub-pixel measurement accuracy, and the vibration caused by the 20-N maneuver could be observed. Subsequently, the measurements during the transition from sunshine to eclipse showed that deformation of the solar array paddle arose around the time when the satellite went into eclipse.

We are scheduled to conduct additional measurements of the thermal snap phenomena which occur at the solar array paddles of GOSAT because we have not succeeded in conducting an uncut thermal snap measurement from sunshine to umbra.

8. References

Foster, C. L.; Tinker, G. S.; Nurre, W. & Till, W. A. (1995). NASA Technical Paper, *The Solar Array-Induced Disturbance of the Hubble Space Telescope Pointing System,* 3556

Iwata, T.; Hoshino, H.; Yoshizawa, T.; Tanamachi, T.; Kawahara, T. & Gonda, H. (2006). Precision Attitude Determination and Control for the Advanced Land Observing Satellite (ALOS): Flight Results (in Japanese), *Proceedings of the 50th Space Science and Technology Conference,* Japan, 2006

Japan Society of Mechanical Engineering (February 2007). *Mechanical Engineers' Handbook Applications γ11: Space Equipment and Systems,* Japan Society of Mechanical Engineering, ISBN 978-4-88898-154-5, Japan

Johnston, J. D. & Thornton, E. A. (2000). Journal of Spacecraft and Rockets, *Thermally Induced Dynamics of Satellite Solar Panel,* Vol. 37, No. 5 (2000) pp. 604-613

Mobara, M. (October 1994). *Introduction to aerospace engineering –guidance and control of satellite and rocket,* Baifukan, ISBN 978-4-56303-493-1, Japan

Part 3

Modelling and Analysis of Complex Systems

8

Modelling and Vibration Analysis of Some Complex Mechanical Systems

Tadeusz Markowski[1], Stanisław Noga[1] and Stanisław Rudy[2]
[1]*Rzeszów University of Technology*
[2]*WSK PZL Rzeszów S.A.*
Poland

1. Introduction

Development of modern engineering requires technical equipment which is characterized by high stability and operational dependability. This particularly applies to components and assemblies which are employed in the aviation, pharmaceutical and biotechnological industries, and biomedical devices. Each newly produced assembly must pass a cycle of advanced computations and a set of experimental tests which allow it to be used in engineering structures. In many cases the experimental tests are realized in order to verify the analytical results. For economic reasons, the calculations should be executed by using advanced computation techniques and the experimental investigations should be conducted on specially designed test rigs. Such test rigs are required to satisfy specific requirements to guarantee that experiments can be led with high technical quality, and economic reasons demand that the rigs should operate for a long period of time (Friswell & Mottershead, 1995). One of the essential factors, which could disturb or restrict the functioning of devices (rigs and others), are the vibration of the components or assemblies of these systems. So it is advantageous to conduct numerical computations at the design stage, which could restrict the consequences connected with the structure vibration. The rapid development of computer techniques and analytical systems based on the finite element method (FEM) allows a free vibration analysis of large systems of complex design and geometry to be conducted (De Silva, 2005). Some authors, using FEM techniques, analyzed the free vibration of complex structures, which are used as load - bearing elements of building structures (Ansell, 2005). Free vibration analysis of selected steel bridges is conducted in papers of Chung and Živanovič (Chung & Sotelino, 2006; Živanovič et al., 2007). The bridge FE models that are developed are verified on real objects. Zembaty (Zembaty et al., 2006) studied the influence of damage of the base construction frame on the change in the natural frequencies of the system. The gas turbine blade is a very important part of aviation jet engine. Some authors analyzed the influence of friction contact phenomenon in collaboration regions of blades and disk sectors on the dynamic behaviour of the blades (Allara, 2009; Toufine et al., 1999). Some presented results were verified by experimental tests. Free vibration of the blade are developed by Sinha (Sinha & Turner, 2011) using the thin shell theory. The achieved solution includes the effect of warping of the cross - section of the blade. Membrane systems have wide application in different disciplines of engineering. Jaffrin and Tack in their papers (Jaffrin, 2008; Tack et al., 2006) present practical examples of the use of the membrane complex systems in the pharmaceutical, chemical and biotechnological industries, and biomedical devices. The fundamental theory of vibration of membrane systems is elaborated in a number of monographs for example by, Kaliski (Kaliski, 1966), Rao (Rao, 2007), and others. The classical theory of membranes is used by Rossit (Rossit et al., 1998) to solve the problem of the free vibration of composite membranes with discontinuously varying density. Noga (Noga, 2008) presents introductory studies that deal with the updating of the FE model of an annular membrane based on the analytical solution data. The complete analytical solutions of undamped free vibration of an elastically connected annular and circular double membrane compound systems are derived by Noga (Noga, 2010a).

The aim of this study is the analysis of the free vibration of selected mechanical systems with complex design and geometry. The present work is organized as follows. Section two gives basic overview of finite element technique for free vibration analysis. Section three presents the vibration problem of the fatigue test rig for aviation gear boxes. In section four the free vibration of the gas turbine blade are analyzed. Section five deals with the vibration problem of an annular membrane resting on an elastic foundation of a Winkler type. Section six gives some concluding remarks.

2. Finite element technique in vibration problem

The finite element (FE) technique is a useful tool to solve various dynamics problems. As mentioned earlier, the use of FE method allows the specific design feature and complex geometry of the system under study to be taken into consideration. The first step in any finite element analysis is to discretize the continuous models by building a discrete model of the physical structure to be analysed. It leads to the equations of motion of the analysed system that may be written in the following form (De Silva, 2005):

$$\mathbf{M}\ddot{\mathbf{u}} + \mathbf{K}\mathbf{u} = 0 \tag{1}$$

where $\mathbf{M}$ and $\mathbf{K}$ are, the global mass and global stiffness matrices (made up by proper assembly of the element matrices); $\ddot{\mathbf{u}}$ and $\mathbf{u}$ are the nodal acceleration and nodal displacement vector, respectively. Global mass and stiffness matrices are assembled from the element matrices that are given by (De Silva, 2005):

$$\mathbf{M}^{(e)} = \int_{V^{(e)}} \rho^{(e)} \mathbf{N}^T \mathbf{N} \, dV^{(e)}, \quad \mathbf{K}^{(e)} = \int_{V^{(e)}} \mathbf{B}^T \mathbf{E} \mathbf{B} \, dV^{(e)} \tag{2}$$

where $\rho^{(e)}$ is the mass density coefficient for an element e; $V^{(e)}$ is the volume of an element e; $\mathbf{N}$ is the matrix of the element shape functions; $\mathbf{B}$ and $\mathbf{E}$ are, the element shape function derivatives, and the elasticity matrices, respectively. The natural frequencies of the system may be obtained by solving the eigenvalue problem

$$\left(\mathbf{K} - \omega^2 \mathbf{M}\right) \bar{\mathbf{u}} = 0 \tag{3}$$

where ω is the natural frequency and $\bar{\mathbf{u}}$ is the corresponding mode shape vector which can be obtained from equation (3). The number of eigenpairs ($\omega_i, \bar{\mathbf{u}}_i$) corresponds to the number of degree of freedom of the system. The block Lanczos method is employed to solve the eigenvalue problem (3). Because of the discretization process, the FE models are treated as approximations of the exact systems. The error between the accurate and the FE models is defined by (Friswell & Mottershead, 1995):

$$\varepsilon = \left(\omega^f - \omega^e\right) / \omega^e \times 100[\%] \tag{4}$$

where ω^f is the natural frequency from the FE analysis, while ω^e is the natural frequency of the exact system. Equation (4) is the so - called frequency error (Friswell & Mottershead, 1995). For many systems with simple geometry, the exact value of the natural frequencies may be achieved from an analytical solution of the free vibration problem (Noga, 2010a, 2010b). Another possibility to determine the exact values of the natural frequencies is the execution of experimental investigation.

3. Vibration of the fatigue test rig for aviation gearboxes

The subject of this section is to analyse the free vibrations of a gear fatigue test rig working in a power circulating rig configuration. Markowski (Markowski et al., 2010) presents introductory studies that deal with the vibration of the system being studied. For this rig two FE models in the ANSYS system are made.

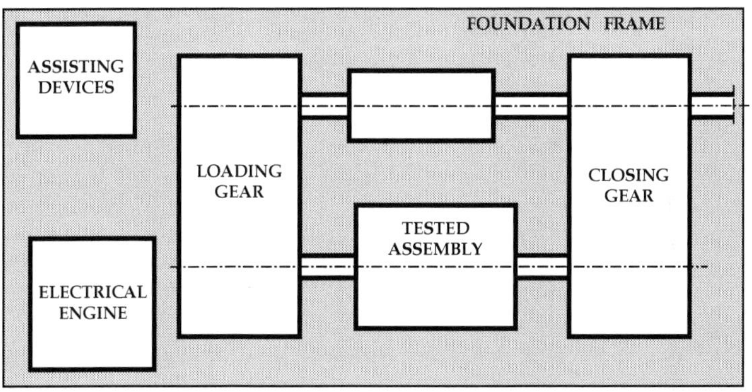

Fig. 1. The block diagram of the test rig

Based on the developed models, natural frequencies and mode shapes are determined. The received results are verified by the experiment. Some concluding remarks are provided.

3.1 Description of the gear test rig working in closed loop configuration

The test rig in question is created at the Test Bed Shop of the Wytwórnia Sprzętu Komunikacyjnego PZL Rzeszów S.A. A block diagram is presented in Fig. 1. All assemblies of the rig are mounted on metallic tables seated on the base frame, the design of which is demonstrated in Fig. 2. The frame consists of two layers of structural sections with a channel section welded together and fastened to the concrete base by anchors. The estimated weight of the base frame is 1600 [kg]. The full description of the base frame can be found in Markowski's paper (Markowski et al., 2010).

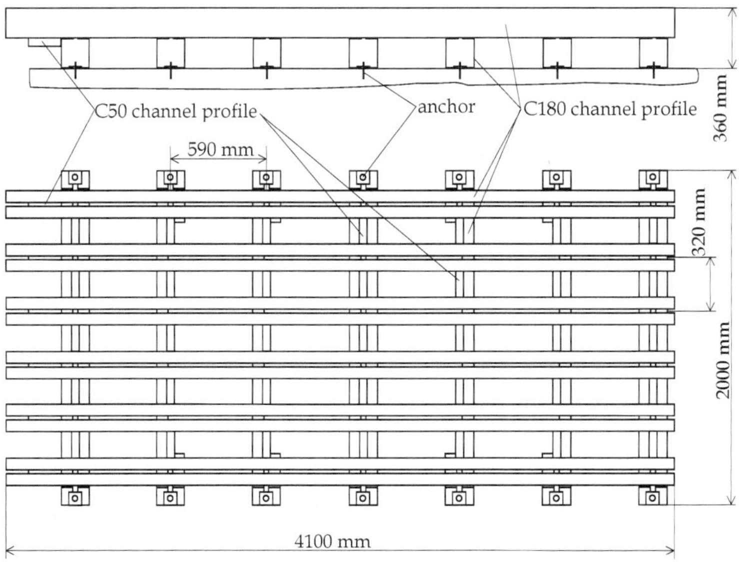

Fig. 2. Model of the base frame

Each additional steel table consisting of a structural 100x100x6 square section or a rectangular 100x50x8 and 100x50x6 sections has a plate made of steel 20 [mm] thick sheet on the upper end. All steel tables of the rig are welded. On the individual plates of the tables, the applicable assemblies of the rig are mounted together with additional features, used for the stability and accuracy of the rig assembly process. In Fig. 3, the rig under investigation is presented. The total estimated weight of the whole rig is 6500 [kg]. The rotating components of the rig are working in a rotating range of 2700 - 7500 [rpm], which, if converted into number of cycles, equals 45 - 125 [Hz] without considering the multiplication factor of the number of teeth. It was found that the assemblied rig vibrates under the influence of ground excitation, which is caused by the work of the neighboring test rigs. There was a real danger that those vibrations would disturb the rig operation.

Fig. 3. Gear fatigue test rig overall view

The necessity of the dynamic rig analysis was recognised in order to estimate its natural frequencies and mode shapes.

3.2 Finite element representations of the system

To obtain satisfactory modal analysis results, an accurate FE model of the system must be developed before conducting an experimental investigation on a real object. Such a primary dynamic overview of the system is helpful in planning and conducting the experimental investigation (De Silva, 2005; Friswell & Mottershead, 1995). The investigated test rig is a complex system composed of various structural components. Therefore the primary aim is to develop an analytical model of the rig that has dynamic parameters (natural frequency and mode shapes) as close as possible to the parameters of a real object. From the economic point of view, the elaborated model should be so simple that the costs connected with its development and operations are kept at a bare minimum (Friswell & Mottershead, 1995). Due to this fact, it is of high importance how crucial structural elements and rig assemblies are modelled, as well as, how proper FE elements are selected. From the test rig stability point of view, the base frame is an essential structural component (Fig. 2). This requires developing of a relatively precise FE model of the base frame. Two finite element models of the investigated system are prepared and discussed with the use of ANSYS FE code.

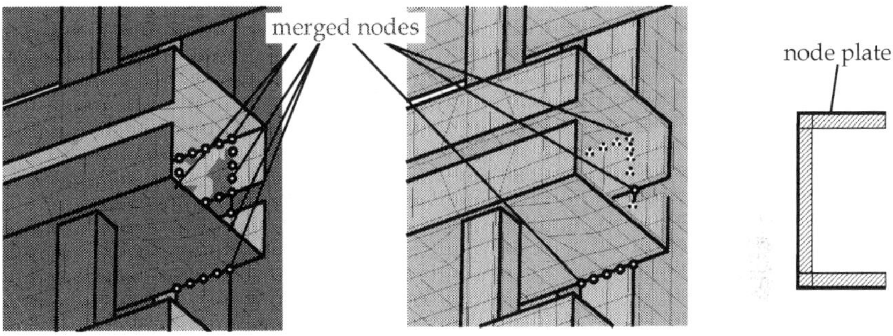

Fig. 4. Modelling of the welded joints

The first FE model is created in the following way: the base frame is modelled using shell elements (*shell99*) quadrilateral, eight nodes with six degrees of freedom on each node and the sets of channel sections of the base frame are modelled with *shell99*, assuming the location of the nodes on the external surface of the element (Fig. 4). However, this approach introduces a modelling error because of the excessive mass concentration in the corner of the channel sections, which results in improper inertia moments of the channel sections sets. The frame support was modelled assuming the constraint point wise in the areas where the frame is attached to the concrete foundation by the anchors. In each such point, one rotational degree of freedom along the axes of the anchor is left free (Fig. 5).

Additional reinforcement of the external sets of the channel sections in the upper layer is modelled by the beam element (*beam44*). Each of the rig assemblies together with corresponding steel tables (with the exception of supporting device) is modelled as a so - called rigid region concentrated at the mass point where total mass is located. In the rigid region, two types of nodes can be distinguished: master nodes and slave nodes subordinated to the master node (Markowski et al., 2010).

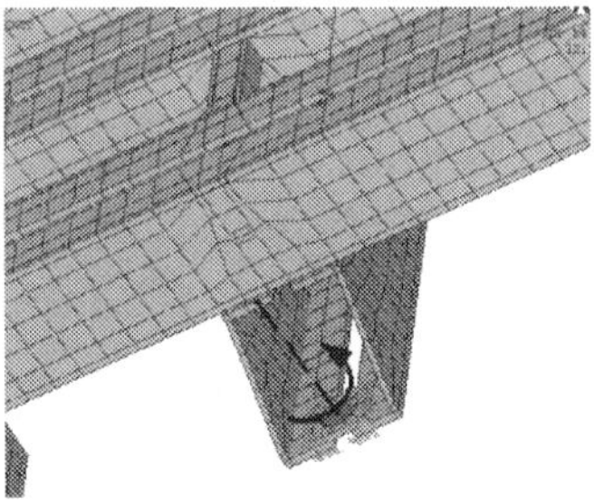

Fig. 5. The boundary condition

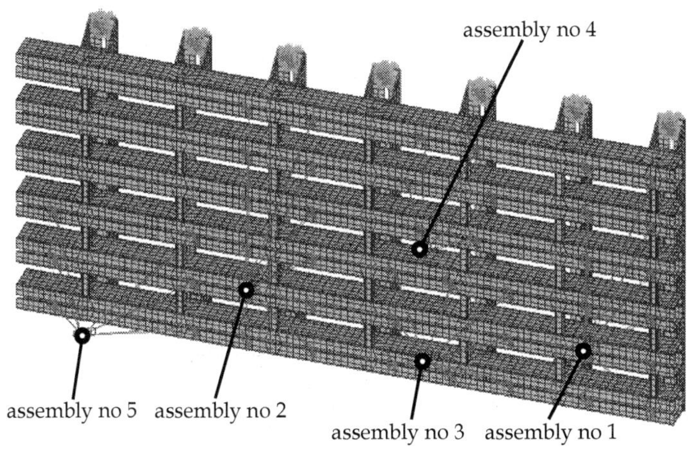

Fig. 6. First FE model of the system

Fig. 7. Finite element model of the assembly no. 1 steel table

This means that slave nodes have the same DOF as the corresponding master node. With reference to the discussed rig, in all the applied rigid regions the location of the master nodes overlaps with the position of mass points located in the centre of gravity of modelled assemblies. Each mass point is realized by a single node mass element (*mass21*) with six DOF. The nodes of the base frame located in the collaboration regions of the modelled assemblies and modelled frame are considered the slave nodes of the particular rigid regions. The discussed FE model of the rig is presented in Fig. 6. The model contains 18208 shell elements (*shell99*), 19 mass point elements (*mass21*), 14 beam elements (*beam44*), and 56892 nodes. As was mentioned earlier, a model of the supporting device is not taken into consideration in this FE model.

In the second FE model case of the rig the base frame is modelled as in the previous case, but the modelling of the assemblies together with the corresponding steel tables are different. In this case the design features of the steel tables of the base frame and mutual connections between individual assemblies are considered. All of that creates a so - called power circulating rig. The bearing elements of each individual table are modelled by a beam element (*beam44*), whereas the steal plates are modelled by a shell element (*shell99*). The required connections and welded joints of each individual steal table component are realized by the node coupling method.

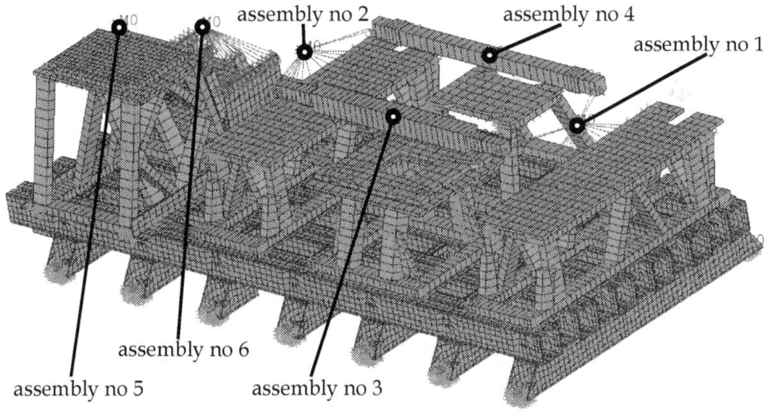

Fig. 8. Second FE model of the system

The assemblies seated on the tables are modelled as a mass point connected by a rigid region to the steel table. As in the previous case, each mass point (*mass21*) is located in the centre of the gravity modelled assembly. The rigid areas of the tables where the modelled assemblies are seated are considered to be the coupled sets of nodes (*"slave" nodes*). The connection of individual tables to the base frame is performed by the coupling function in the clamping areas. In Fig. 7 the table FE model with no. 1 assembly seated on it is presented. The shafts and the clutch assemblies in a power circulating rig are modelled by a beam element (*beam44*) and a spring element (*combin14*), and allow taking into account the elastic properties of the clutch. In the discussed model all important components of the analyzed rig are considered. The developed FE model of the rig consists of 20366 shell99 elements, 1625 beam44 elements, 28 mass21 elements, 12 combine elements and 66026 nodes. The discussed model is shown in Fig. 8.

3.3 Numerical calculations

Numerical analysis results of natural frequencies of the gear fatigue test rig are obtained using the models presented earlier. For each approach, numerical calculations are conduced to evaluate natural frequencies of the system and corresponding mode shapes in the frequency range 0 to 300 [Hz]. For the steel elements used for the rig, the following data materials are used: Poisson ratio $\nu = 0.3$, Young's modulus $E = 2.1*10^{11}$ [Pa], and density $\rho = 7.86*10^3$ [kg/m^3]. The results are split into two categories. The natural frequencies and mode shapes related to the movement of assemblies mounted on the base frame are considered to be the first category. The other category includes the natural frequencies and mode shapes related to the local movement of the channel section sets of the base frame. The vibration of the assemblies with mode shapes, which could be considered the first category, have greater consequences for a proper rig operation because of their movement when the rig is running. The vibration of these particular assemblies can be realized as a concurrent oscillation form when the sense of movement has the same signs or a backward oscillation form when sense of movement has the opposite signs. For both FE models, numerical analysis results are presented with reference to the movements of the particular rig assemblies. In order to unambiguously describe the mode shapes presented, it is assumed that longitudinal movement is a movement in the plane parallel to the base and along the longer side of the base frame (Fig. 2). Transverse movement is a movement in the plane parallel to the base and along the shorter side of the base frame. The vertical movement is considered perpendicular to the base movement. For both FE models, the discussed results are presented in the sequence of appearance. At first the results generated from the first rig FE model are presented. The mass of the assemblies and the supporting tables (Fig. 6) required for the analysis is presented in Tab. 1.

Assembly no.	1	2	3	4	5
mass [kg]	1480	1480	550	320	600

Table 1. Evaluated mass of the particular assemblies of the test rig (first FE model)

The obtained natural frequency results and the description of related modes are included in Tab. 2. The graphic presentations of the discussed results are shown in Fig. 9 - 12.

Mode no.	Mode shape description	Value of the natural frequency ω^f [Hz]	Figure no.
P1	Concurrent longitudinal vibration all assemblies	22.068	9a
P2	Vertical vibration of the assembly no. 1	37.764	9b
P3	Backward vertical vibration of the assembly no. 1 and 2	40.661	9c
P4	Concurrent transverse vibration all assemblies	51.745	10a
P5	Vertical vibration of the assembly no. 5	57.672	10b
P6	Backward transverse vibration assemblies no. 1 and 2	59.273	10c
P7	Vertical vibration of the assembly no. 3	73.346	11a
P8	Vertical vibration of the assembly no. 4	85.160	11b
P9	Concurrent transverse vibration assemblies no. 3 and 4 and backward with assembly no. 5	93.031	11c
P10	Concurrent vertical vibration of the assemblies no. 3, 4 and 5	101.70	12a
P13	Concurrent vertical - transverse vibration of the assemblies no. 3 and 4.	138.71	12b

Table 2. Natural frequency and mode shapes of the test rig (first FE model)

The values of the natural frequencies of the test rig corresponding to modes P3 - P10 are within the operating range of the rotating parts of the rig assemblies.

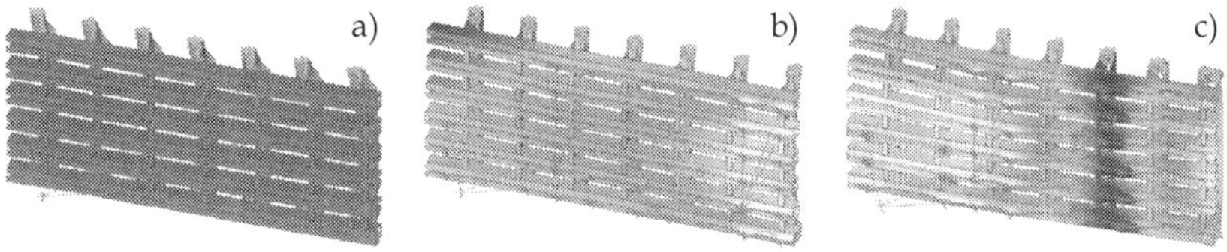

Fig. 9. Mode shapes: (a) P1, (b) P2, (c) P3

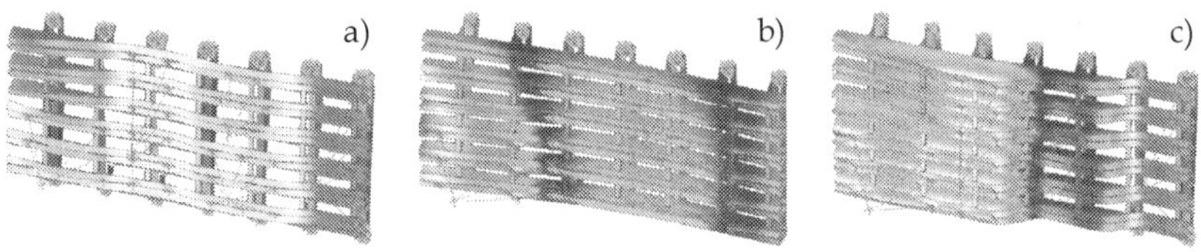

Fig. 10. Mode shapes: (a) P4, (b) P5, (c) P6

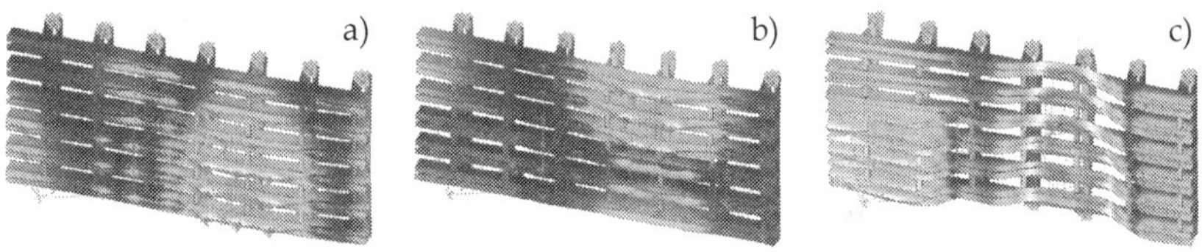

Fig. 11. Mode shapes: (a) P7, (b) P8, (c) P9

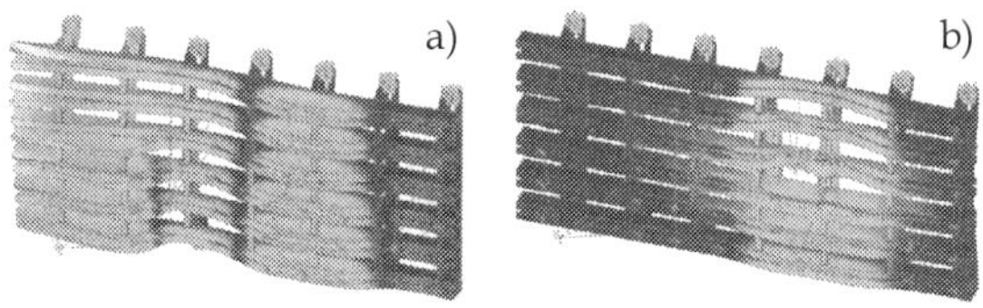

Fig. 12. Mode shapes: (a) P10, (b) P13

Subsequently the results of the second FE model of the rig are obtained as shown in Fig. 8. As mentioned before the design features of the tables located under the assemblies of the rig and the connections between the individual assemblies are taken under consideration, and those created as so - called a power circulating rig that creates a so - called a power circulating rig. Included in the calculations are the estimated masses of particular assemblies shown in Fig. 8, which are modelled by rigid, which mass are value presented in Tab. 3.

Assembly no.	1	2	3	4	5	6
mass [kg]	1100	1200	350	150	290	320

Table 3. Evaluated mass of the particular assemblies of the test rig (second FE model)

The received frequencies with their corresponding description are shown in Tab. 4a - b. Figs. 13 - 20 presents the discussed mode shapes.
As expected, based on the second FE model of the rig, a greater number of natural frequencies and corresponding modes are received in comparisons to the first FE model. Moreover the consideration of the design features of the tables allowed for more accurate results pertaining to the range of form of particular natural frequencies. The values of the natural frequencies of the rig corresponding to modes D5 - D19 are within the operating range of the rotating parts of the rig assemblies. From the analysis of the received vibration forms it can be concluded that the tables seated on the base frame supporting assemblies are practically not subjected to deformation (they are characterized by higher stiffness in relation to the base frame). Some part of the received results is characterized by a qualitative similarity to the majority of the solutions received from the first rig FE model. A qualitative similarity between forms D2 and P1, D4 and P2, D5 and P3, D6 and P4, D10 and P6, D16 and P7, D18 and P8, D24 and P13 can be observed. A similarity to the solution from the second model is not observed for forms P5, P9, P10 from the first FE model.

Mode no.	Mode shape description	Value of the natural frequency ω^f [Hz]	Figure no.
D1	Longitudinal vibration of the assembly no. 6	14.01	13a
D2	Concurrent longitudinal vibration all assemblies	21.07	13b
D3	Vertical vibration of the assembly no. 6	23.15	13c
D4	Concurrent vertical vibration of all assemblies besides assembly no. 6. Additionally swinging transverse backward motion assemblies no. 3 and 4	38.49	14a
D5	Swinging longitudinal motion of assembly no. 4 and vertical backward vibration of assembly no. 1 against assembly no. 2 and 5	41.21	14b
D6	Transverse concurrent vibration all assemblies and vertical backward vibration of assembly no. 1 against assembly no. 2 and 5	41.52	14c
D7	Swinging longitudinal motion of assembly no. 4 and vertical backward vibration of assembly no. 1 and 6 against assembly no. 2 and 5	43.77	15a
D8	Transverse backward vibration of assembly no. 1, 4 and 3 against assembly no. 2, 5 and 6	45.65	15b
D9	Swinging longitudinal motion of assembly no. 4 and 5, backward vibration of assembly no. 1, 2 and 3	48.53	15c

Table 4a. Natural frequency and mode shapes of the test rig (second FE model)

Mode no.	Mode shape description	Value of the natural frequency ω^f [Hz]	Figure no.
D10	Transverse backward vibration of assembly no. 1, 5 against assembly no. 2, 3 and 4	52.42	16a
D11	Swinging transverse backward motion of assemblies no. 3 and 4 and longitudinal vibration of assembly no. 5	54.34	16b
D12	Transverse backward vibration of assembly no. 1 and 2 against assemblies no. 3, 4 and 5	55.91	16c
D13	Swinging longitudinal vibration of assembly no. 5 and transverse motion of the assembly no. 4	59.35	17a
D14	Dominant swinging transverse motion of assembly no. 3 and transverse vibration of assembly no. 6	68.78	17b
D15	Transverse vibration of assembly no. 6, and swinging motion of assembly no. 3	69.46	17c
D16	Vertical backward vibration of assemblies no. 3 and 5 against assemblies no. 1 and 2	71.26	18a
D17	Vertical backward vibration of assemblies no. 3 and 5 and longitudinal backward vibration of assemblies no. 1 and 2	79.69	18b
D18	Vertical vibration of assembly no. 4	94.84	18c
D19	Vertical vibration of the base frame under assembly no. 6	112.78	19a
D20	Longitudinal concurrent vibration all assemblies (second form, mass points are immovable)	129.12	19b
D24	Transverse vibration of the base frame under assemblies no. 3 and 4 (mass points are motionless)	155.83	19c
D31	Transverse vibration of the base frame under assemblies no. 5 and 6	165.85	20

Table 4b.Natural frequency and mode shapes of the test rig (second FE model)

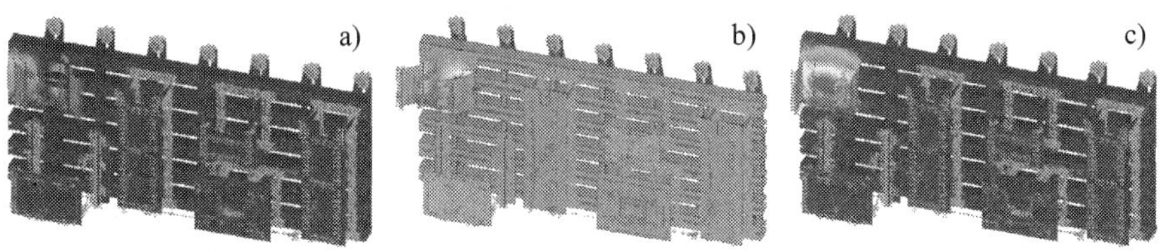

Fig. 13. Mode shapes: (a) D1, (b) D2, (c) D3

Fig. 14. Mode shapes: (a) D4, (b) D5, (c) D6

Fig. 15. Mode shapes: (a) D7, (b) D8, (c) D9

Fig. 16. Mode shapes: (a) D10, (b) D11, (c) D12

Fig. 17. Mode shapes: (a) D13, (b) D14, (c) D15

Fig. 18. Mode shapes: (a) D16, (b) D17, (c) D18

Fig. 19. Mode shapes: (a) D19, (b) D20, (c) D24

Fig. 20. Mode shape D31

3.4 Experimental investigations

The prepared FE models of the test rig are verified by the experimental investigation on a real object (Fig. 3). A Brüel and Kjær measuring set is used in the experimental investigation.

Fig. 21. The measuring test

The set consisted of the 8202 type modal hammer equipped with a gauging point made of a composite material, the 4384 model of accelerometer, the analogue signal conditioning system, the acquisition system, and the data processing system supported by Lab View analytical software. The analysis of the results of the experimental investigation is conducted on a portable computer using actual measured values. The measurement experiment is scheduled and conducted to identify natural frequencies and corresponding mode shapes related to the transverse, longitudinal and vertical vibration of the assemblies no. 1 and 2, respectively. Because only one accelerometer was accessible, the measurement

process is conducted in a so - called measurement group. For each group, the accelerometer position for a tap place point for the hammer (impulse excitation) is established.
When the location of the measurement points for a particular group was to be determined a numerical calculation was used as reference. The experiments are planned and conducted for five measurement groups. The first group is made up of points 1 to 6, and is located on the base frame and table no. 1 (Fig. 22). The accelerometer is located in point no. 2. The second group consists of points 7, 8, 9, 10 and 14, and is located on the table of assemblies no. 1, 2, 3 and 5 (Fig. 22 and 23).

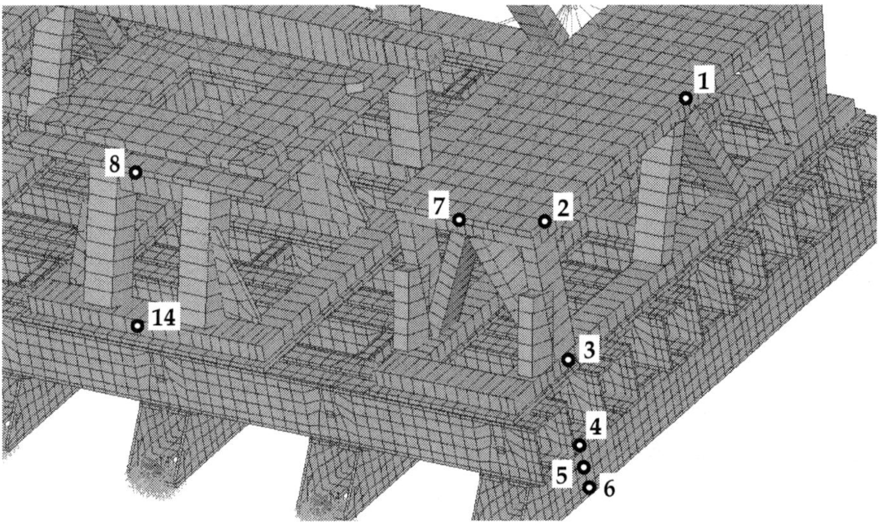

Fig. 22. Measuring set points

Mode no.	Measuring set no.	Measured natural frequency value ω^e [Hz]	Frequency relative error ε [%]
P1	1, 4	27.77	-20.5
P2	1	38.15	-1.01
P4	2	46.70	10.8
P7	2	73.24	0.15

Table 5. Experimental investigation results related to the first FE model

The accelerometer for this group is located in point no. 8. The third measurement group is made up of points no. 10 and 11 (Fig. 23), while the experiment is conducted the accelerometer is located in point no. 10 and subsequently in point no. 11. The fourth measurement point is made of points 13 and 15, and the accelerometer is located in point no. 13 (Fig. 23). The fifth group consists of points 12 and 16, where the accelerometer was located in point 12. For all the discussed cases the impulse response is registered which

caused modal hammer vibrations in each of the mentioned points. Tables 5 and 6 present the natural frequencies excited and identified in the measurement experiment, their corresponding mode shapes, and frequency error defined according to formula (4).
The results presented in Tab. 5 refer to the first FE model of the system, whereas the results for the second FE model are shown in Tab. 6.
Identification of the form is conducted by a qualitative comparison of the numerical and experimental results. In Fig. 24, the frequency characteristic of the system for the first measured group is presented. Fig. 24a presents the amplitude - phase characteristic, whereas Fig. 24b presents the phase - frequency characteristic.

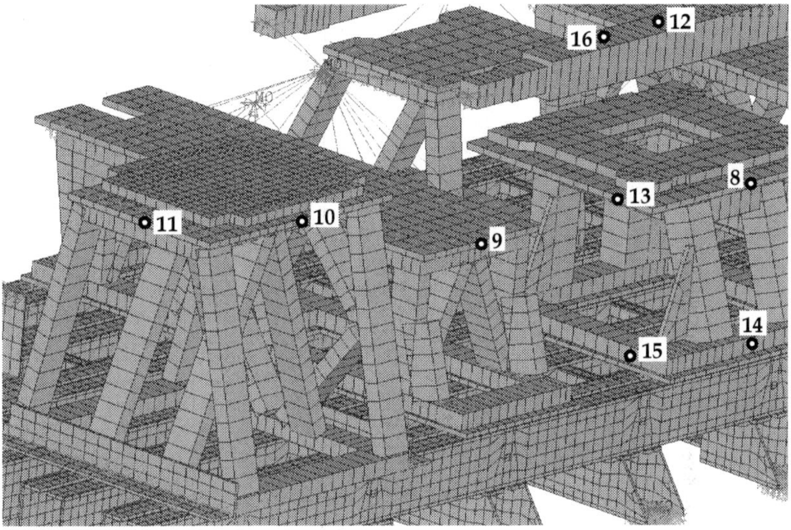

Fig. 23. Measuring set points

Mode no.	Measuring set no.	Measured natural frequency value ω^e [Hz]	Frequency relative error ε [%]
D2	1, 4	27.77	-24.1
D4	1	38.15	0.9
D6	2	46.70	-11.1
D9	1	50.05	-3.0
	3	50.35	-3.6
D11	2	55.85	-2.7
	3	56.15	-3.2
D13	5	61.34	-3.2
D14	4	65.30	5.3
D16	2	73.24	-2.7
D24	2	153.20	1.7

Table 6. Experimental investigation results related to the second FE model

When analyzing the received results (Tab. 5 and 6), a small difference can be observed in both cases between the numerical results and the experiment related to the frequencies connected with the vertical vibration (forms P2 and P7 of the first FE model and D4 and D16 of the second one). A relatively small difference can be observed for natural frequencies related to the complex forms of vibration where there is a combination of vertical and transverse vibration or transverse vibration of the assemblies no. 3 and 4 (forms D9, D11, D13, D14, D16 and D24 of the second FE model). For both models significant differences occur for the natural frequencies connected to the concurrent vibrations in the base frame plane of the rig (forms P1 and P4 of the first FE model and forms D2 and D6 of the second FE model).

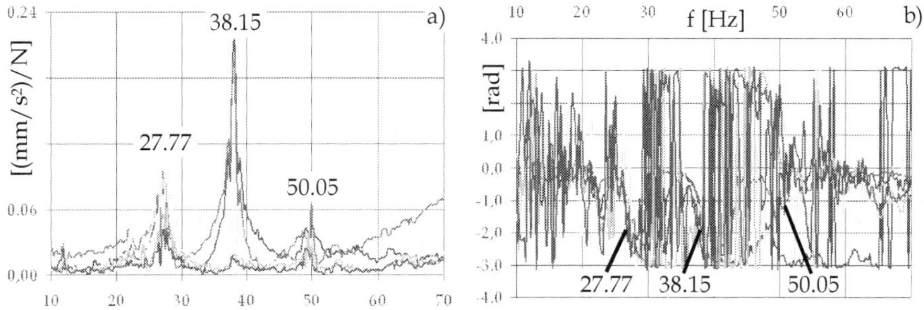

Fig. 24. Frequency characteristic of the system

4. Vibration of the aviation engine turbine blade

In this section, the free vibration of an aviation engine turbine blade is analyzed. Rudy (Rudy & Kowalski, 1998) presents the introductory studies connected with the discussed problem. In the elaborated blade FE models a complex geometrical shape and the manner of the blade attachment to the disk are taken into consideration. Some numerical results are verified by the measurement experiment.

4.1 Free vibration of the engine turbine blades

Gas turbine blades are one of the most important parts among all engine parts. Those elements are characterized by complex geometry and variations of material properties connected with temperature. Moreover, it is necessary to take into account the manner of the blade attachment to the disk. The most popular is fixing by a so - called fir tree. During operation the blade vibrates in different directions. To facilitate consideration circumferential, axial and torsional vibration are distinguished but as a matter of fact circumferential and axial vibration are bending. In fact all mentioned vibrations are a compound of torsional and bending vibrations. Each vibrating continuous system is described by unlimited degrees of freedom and consequently unlimited number of natural frequencies. The blade vibration with the lowest value is called the first order tangential mode. For the analytical calculation of natural frequencies of a blade, the usual assumption is that of the Euler - Bernoulli model of the beam (Łączkowski, 1974) with constant cross - section fixed in one end. There is significant variability of geometrical parameters long ways of the blade. In accordance with the mentioned approach, for the blade with geometrical parameters at the bottomsection, the natural frequency is equal to 1189.2 [Hz] and for geometrical parameters of the midspan of the blade the natural frequency is equal to 931.6 [Hz]. To achieve more accurate results, the more precise model which takes into account the variability of blade geometry and the manner of the blade fixing to the disk, needs to be prepared.

4.2 Conception of the engine turbine blade finite element formulation

The prepared FE model consists of a sector of the disk with the blade. To decrease the complexity of the model, a sector of the disk with an angle resulting from the number of assembled blades is limited to a specific radius at the bottom side. The blade and the disk sector are modeled with the use of solid elements. In most cases hexahedral elements are used, however, the model consists of wedge and tetrahedral elements as well. In the analyzed case, blades are attached to the disk by a fir tree with three lobes. The collaboration regions of the blade and the sector disk are modelled by using the 3D contact elements. Those elements allow taking into account the relative displacement of the faces in contact under the influence of an external load. A slip soft contact element is used with a friction coefficient equal 0.1. Because the prepared model includes only the circular sector of the disk, it is necessary to apply cyclic symmetry boundary conditions. Moreover, because the disk is limited to a specific internal diameter it is necessary to apply a proper DR displacement to model the removed part of the disk. The DR value is initially calculated using an axisymmetrical model where blade load was modelled as uniform pressure on the rim of the disk.

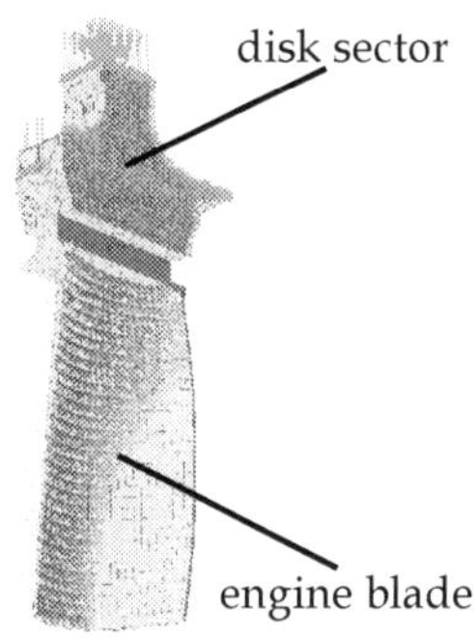

Fig. 25. Finite element model of the system

The discussed FE model of the blade with the sector disk is shown in Fig. 25. The blade is modelled by using 7986 solid elements and it has 10155 nodes. The disk sector is modelled by using 9646 solid elements with 12056 nodes. The contact region include 210 3D contact elements. The FE model in question is performed in MSC/PATRAN system, whereas dynamic analysis is performed in MSC/ADVANCED_FEA solver.

4.3 Numerical calculations

Numerical analysis of the engine turbine blade with the disk sector free vibration is obtained using the model suggested earlier. For each approach, only the first nine natural frequencies

and mode shapes are evaluated. For the blade, the following data materials are used: Poisson ratio ν = 0.3, Young's modulus E = $2.1*10^{11}$ [Pa], and density ρ = $8.2*10^{3}$ [kg/m^3]. Specially, the impact of the manner of the blade fixing to the disk in the FE models on the value of the natural frequencies of the blade is analyzed. At first, analysis for nominal dimensions of the fir tree is performed (the load is distributed uniformly on each lobe). Then, two extreme cases are chosen, where lobes are made in such a way that the top lobes are loaded more in one case and the bottom lobes in another. Tolerance limit for analysis is assumed at the level of 0.02 [mm]. The forces generated the lobes loading derive from the centrifugal forces arisen within blade during its rotation. For these cases the calculation are executed assuming that the systems rotate with the rotational speed equal to 15100 [rpm].
For the next analyzed instance the natural frequencies of the engine blade without the disk sector are computed. It is assumed that the blade is fixed on contact faces of the lobe and the rotational speed is equal to 0, 7550 [rpm] and 15100 [rpm], respectively. Frequency results for such models are presented in Table 7.

Mode no.	Value of the natural frequency ω^f [Hz]					
	Blade without the disk sector			Blade with the disk sector		
	0 [rpm]	7550 [rpm]	15100 [rpm]	Nom. dim. of fir tree	Top lobe more loaded	Bottom lobe more loaded
1	1134.0	1177.2	1260.6	1124.8	1141.4	1069.7
2	2251.9	2290.2	2320.0	1629.3	1635.7	1603.3
3	3705.0	3720.5	3636.4	2524.0	2533.1	2495.0
4	5182.0	5230.0	5184.8	4021.1	4067.3	3871.6
5	6510.0	6530.4	6361.3	5412.8	5476.6	5238.5
6	9218.0	9256.5	9031.5	7548.4	7598.3	7332.1
7	9945.9	9969.7	9689.1	8782.1	8863.5	8379.8
8	11125	11170	10906	9633.2	9709.6	9277.1
9	14174	14217	13827	9846.8	10046	9672.0

Table 7. Natural frequencies of the system under study

First three mode shapes of vibration corresponding to the presented pairs of the natural frequencies are presented in Fig. 26.

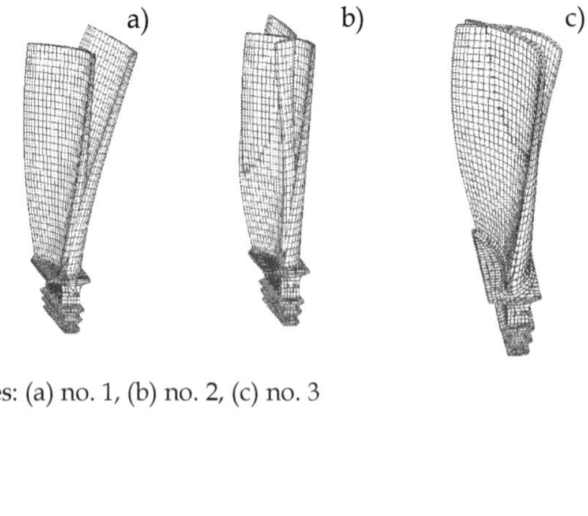

Fig. 26. Mode shapes: (a) no. 1, (b) no. 2, (c) no. 3

It can be seen that performing the lobes on the blade and disk have an influence on the value of blade natural frequencies. For example when the top lobe of the fir tree is more loaded, the first natural frequency increases by 1.5 [%] while in the case that the bottom lobe is more loaded, frequency decreases by 4,9 [%] in comparison to the nominal blade configuration. In the instance where the blade is fixed on contact faces of the lobe, the natural frequencies are higher as in the case of the blade with the disk sector. For example, natural frequencies for 1, 2 and 3 form increase correspondingly by 12,1 [%], 42,4 [%] and 44,1 [%] in relation to blade with the disk sector with nominal configuration.

Usually, endurance tests of blades are conducted on a shaker table. Such experimental investigations are useful for verification the proposed FE model of the blade. The blade is mounted in fixing part of the circular section of the disk and it is excited with the range of frequency 1020 - 1120 [Hz], which refers to the first natural frequencies. A bit lower few percent value of natural frequency received during the test is as a result of flexibility of the fixture seated on the shaker.

5. Vibration of the annular membrane resting on an elastic Winkler – type foundation

In this section, the free transversal vibration of the annular membrane attached to a Winkler foundation is studied using analytical methods and numerical simulation. The introductory studies related to this problem are presented by Noga (Noga, 2010b). At first the general solution of the free vibration problem is derived by the Bernoulli - Fourier method. The second model is formulated by using finite element representations. The results received from the analytical solution (natural frequencies and its mode shapes) allow the determination of the quality of the developed FE models.

5.1 Theoretical formulation

The mechanical model of the system under study consists of an annular membrane resting on a massless, linear, elastic foundation of a Winkler type. It is assumed that the membrane is thin, homogeneous and perfectly elastic, and it has constant thickness. The membrane is uniformly stressed by adequate constant tensions applied at the edges (Fig. 27).

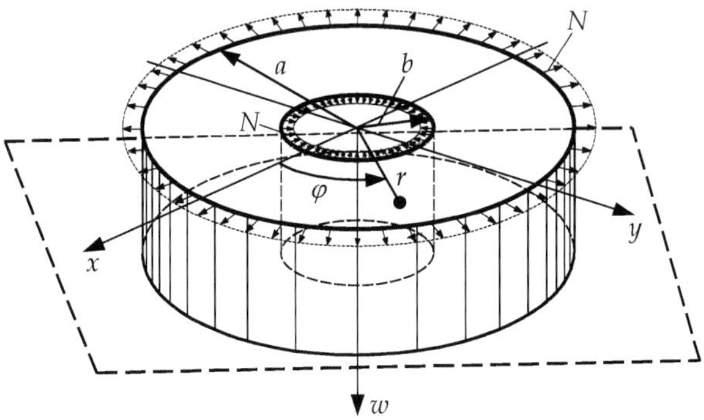

Fig. 27. Vibrating system under study

Making use of the classical theory of vibrating membranes, the partial differential equations of motion for the free transversal vibrations are

$$m_D \ddot{w} - N\,\Delta w + k w = 0 \tag{5}$$

where $w = w(r,\varphi,t)$ is the transverse membrane displacement, r,φ,t are the polar coordinates and the time, a,b,h are the membrane dimensions, ρ is the mass density, N is the uniform constant tension per unit length, k is the stiffness modulus of a Winkler elastic foundation and

$$m_D = \rho h, \quad \dot{w} = \frac{\partial w}{\partial t}, \quad \Delta w = \frac{\partial^2 w}{\partial r^2} + \frac{1}{r}\frac{\partial w}{\partial r} + \frac{1}{r^2}\frac{\partial^2 w}{\partial \varphi^2} \tag{6}$$

The boundary and periodicity conditions are

$$w(a,\varphi,t) = w(b,\varphi,t) = 0, \quad w(r,\varphi,t) = w(r,\varphi + 2\pi,t) \tag{7}$$

Now using the separation of variables method (Kaliski 1966), one writes

$$w(r,\varphi,t) = W(r,\varphi)T(t), \qquad T(t) = C\sin(\omega t) + D\cos(\omega t) \tag{8}$$

where ω is the natural frequency of the system. Introducing solutions (8) into (5) gives the following expression

$$\Delta W + k_D^2 W = 0, \qquad k_D^2 = \left(m_D\omega^2 - k\right)/N \tag{9}$$

The coefficient k_D^2 is positive when $\omega^2 > k/m_D$. This condition guarantees harmonic free vibrations (Noga 2010). The solution of equation (9) is assumed in the form

$$W(r,\varphi) = R(r)U(\varphi) \tag{10}$$

The boundary and periodicity conditions in terms of $R(r)$ and $U(\varphi)$ become

$$R(a) = R(b) = 0, \qquad U(\varphi) = U(\varphi + 2\pi) \tag{11}$$

Substituting solution (10) into (9) yields

$$R_n(r) = A_n J_n(k_D r) + B_n Y_n(k_D r), \; U_n(\varphi) = L_n \sin(n\varphi) + M_n \cos(n\varphi), \; n = 0,1,2,\ldots \tag{12}$$

where A_n, B_n, L_n, M_n are unknown coefficients and $J_n(\cdot)$, and $Y(\cdot)$ are the first and second kinds of Bessel functions of order n. Conditions (11) yields a system of two linear, homogeneous equations in the constants A_n, B_n. Finally, a determinant gives the equation for the natural frequencies from the non - triviality condition. It yields the secular equation

$$J_n(k_D a)Y_n(k_D b) - J_n(k_D b)Y_n(k_D a) = 0 \tag{13}$$

From the relation (13) it is proved that $k_D = k_{mn}\ (m = 1,2,3,\ldots)$ are the roots of the above equation. Then taking into account equation (9), the natural frequencies of the system under consideration are determined from the relation

$$\omega^2 = \omega_{mn}^2 = \left(k_{mn}^2 N + k\right)/m_D \tag{14}$$

The general solution of the free vibrations of the system under study takes the form

$$\begin{aligned} w(r,\varphi,t) &= \sum_{m=1}^{\infty}\sum_{n=0}^{\infty} W_{mn}(r,\varphi)T_{mn}(t) = \sum_{m=1}^{\infty}\sum_{n=0}^{\infty}\Big(\Big(C_{mn}^{(1)}\sin\left(\omega_{mn}t\right) + D_{mn}^{(1)}\cos\left(\omega_{mn}t\right)\Big)\times \\ &\times W_{mn}^{(1)}(r,\varphi) + \Big(C_{mn}^{(2)}\sin\left(\omega_{mn}t\right) + D_{mn}^{(2)}\cos\left(\omega_{mn}t\right)\Big)W_{mn}^{(2)}(r,\varphi)\Big) \end{aligned} \tag{15}$$

where

$$\begin{aligned} W_{mn}^{(1)}(r,\varphi) &= \left(e_{mn}J_n\left(k_{mn}r\right) + Y_n\left(k_{mn}r\right)\right)\sin(n\varphi) \\ W_{mn}^{(2)}(r,\varphi) &= \left(e_{mn}J_n\left(k_{mn}r\right) + Y_n\left(k_{mn}r\right)\right)\cos(n\varphi) \end{aligned} \tag{16}$$

are two linear - independent mode shapes, and

$$e_{mn} = -Y_n\left(k_{mn}a\right)/J_n\left(k_{mn}a\right) \tag{17}$$

5.2 Finite element representations

Finite element models are formulated to discretize the continuous model given by equation (5). As mentioned earlier, the FE models are treated as an approximation of the exact system. The quality of the approximate model depends on the type and density of the mesh. The essential problem of this section is building the FE model of the elastic foundation. The two FE models with different realizations of the Winkler elastic foundation are prepared and discussed by using ANSYS FE code. The first FE model is realized as follows. The foundation is modeled by a finite number of parallel massless springs. The stiffnes modulus k_S of each spring can be obtained from the relation (Noga, 2010a)

$$k_S = \left(k\,p_0\right)/b \tag{18}$$

where p_0 is the area of the membrane large face and b is the number of the springs. The spring - damper element (*combin14*) defined by two nodes with the option *"3 - D option longitudinal"* is used to realize the elastic foundation. The damping of the element is omitted. The layer consists of 9324 combin elements. The annular membrane is divided into 9540 finite elements. The four node quadrilateral membrane element (*shell63*) with six degree of freedom in each node is used to realize the membrane. It is shown by Noga (Noga 2010) that satisfactory results are achieved by realizing the tensile forces as follows. On each node lying on the outer edge is imposed a concentrated tensile force N_0 in the radial direction. The proper value of the force is selected experimentally by numerical simulation. The boundary conditions are realized as follows. All nodes lying on the outer edge of the membrane are simply supported with a possibility to slide freely in the radial direction, and all nodes lying on the inner edge of the membrane are pinned.

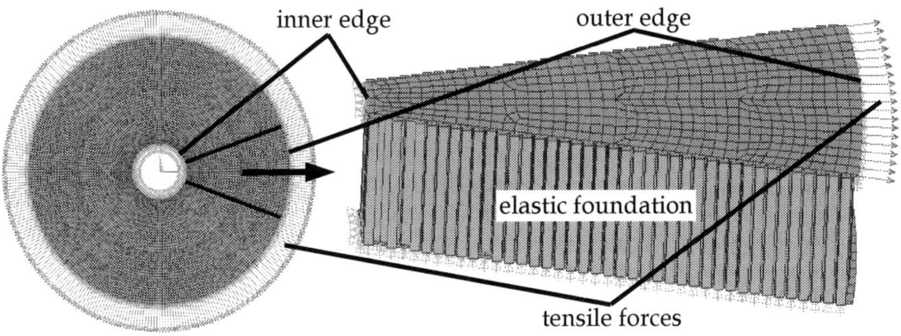

Fig. 28. Second finite element model of the system

The second FE model is the same as the first, but the execution of the Winkler foundation is different. Each massless spring is modeled by using a bar (*link*) element. The values of the dimension parameters of the bars are established a priori. The proper value of the Young's modulus E_f of each bar is selected experimentally to minimize the frequency error (4).

5.3 Numerical analysis

Numerical solutions for free vibration analysis of the annular membrane attached to elastic foundation models suggested earlier are computed. For each approach, only the first ten natural frequencies and mode shapes are discussed. Table 8 presents the parameters characterizing the system under study.

a [m]	b [m]	h [m]	ρ [kg/m^3]	E [Pa]	ν	N *[N/m]*
0.5	0.1	0.002	$2.7 \cdot 10^3$	$7 \cdot 10^{10}$	0.32	500

Table 8. Parameters characterizing the system under study

In the table, E and ν are, the Young's modulus and Poisson ratio, respectively. For the continuous model the natural frequencies are determined from numerical solution of the equations (13) and (14). The results of the calculation are shown in Table 9.

		n					
		0	1	2	3	4	5
m	1	12.0828	13.3304	16.2846	19.8242	23.4497	27.0413
	2	24.0425	24.8625	27.1393	30.3978		

Table 9. Natural frequencies of the system under study ω_{mn}^{e} [Hz] (exact model)

The natural frequencies and the frequency error presented in Tables 10 - 11 are referred to the first FE model and are obtained for $N_0 = 4.8[N]$.

		n					
		0	1	2	3	4	5
m	1	12.486	13.516	16.094	19.377	22.861	26.359
	2	24.85	25.486	27.302	30.046		

Table 10. Natural frequencies of the system under study ω_{mn}^{f} [Hz]

		n					
		0	1	2	3	4	5
m	1	3.337	1.3923	-1.1704	-2.2558	-2.5105	-2.5232
	2	3.3586	2.5078	0.5995	-1.1573		

Table 11. Frequency error ε_{mn} [%]

Tables 12 - 13 show the results related to the second FE model of the system under study and are obtained for $N_0 = 4.8[N]$ and $E_f = 265[Pa]$.

		n					
		0	1	2	3	4	5
m	1	12.485	13.515	16.093	19.376	22.86	26.358
	2	24.849	25.485	27.301	30.045		

Table 12. Natural frequencies of the system under study ω_{mn}^{f} [Hz]

		n					
		0	1	2	3	4	5
m	1	3.3287	1.3848	-1.1766	-2.2609	-2.5147	-2.5269
	2	3.3545	2.5038	0.5958	-1.1606		

Table 13. Frequency error ε_{mn} [%]

For both FE model cases the biggest difference between the exact and the FE solutions may be visible for the frequencies ω_{10} and ω_{20}, respectively. For all cases the sequence of appearance of the natural frequencies refered to the adequate modes shapes is the same. Some modes of vibration corresponding to the presented pairs of natural frequencies are presented in Fig. 29 and Fig. 30, respectively.

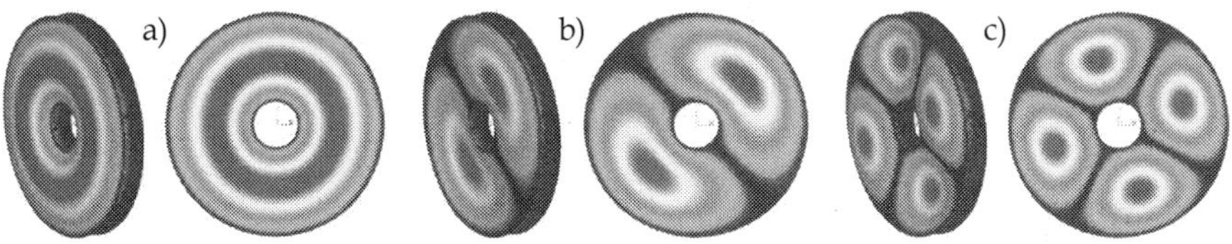

Fig. 29. Mode shapes corresponding to the frequencies: (a) ω_{10}, (b) ω_{11}, (c) ω_{12}

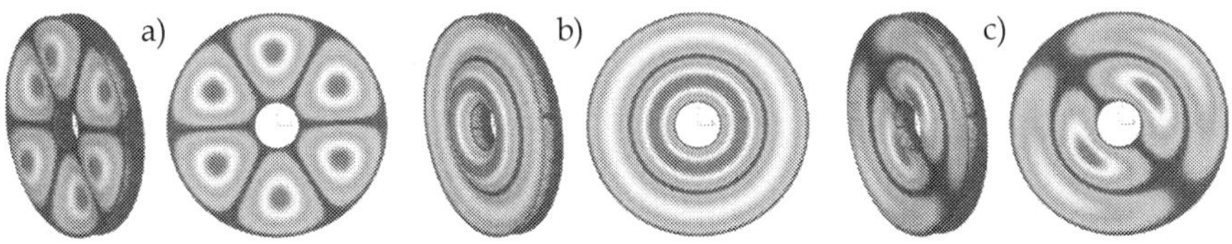

Fig. 30. Mode shapes corresponding to the frequencies: (a) ω_{13}, (b) ω_{20}, (c) ω_{21}

6. Conclusion

This work deals with the analysis of the free vibration of selected mechanical systems with complex design and geometry. Three different types of mechanical systems are taken into consideration, i.e.: a fatigue test rig for aviation gear boxes, a gas turbine blade and an annular membrane attached to Winkler elastic foundation. Design and creation of modern devices require the use of advanced numerical software based on the finite element method. It is specially addressed to modern aviation test rigs and other aviation parts like turbine blades. This allows the specific design features and complex geometry of the rig and blade to be taken into consideration. Two FE models of the test rig are investigated. Some portion of the natural frequencies and mode shapes obtained from numerical calculations are verified with experimental investigations. Considering the obtained results, a small difference can be observed in both cases between the numerical and experimental results related to the natural frequencies connected with the vertical vibration forms of the rig. For both models, lower consistency appears for the natural frequencies connected to the concurrent vibration in the base frame plane of the rig. The second FE model of the rig gives satisfactory insight on the dynamic behaviour of the rig. Further investigation will be oriented towards developing a numerical model allowing better consistency between numerical and experimental results. The FE model of the analyzed blade is verified during an endurance test with accuracy to the first natural frequency. When analyzing the obtained results, it can be seen that performing the lobes on the blade and the disk sector has an influence on the value of blade natural frequencies. Based on the classical theory of membranes, a comprehensive vibration analysis of an annular membrane attached to elastic foundation of the Winkler type is investigated. The Bernoulli - Fourier method is applied to derive the eigenvalue problem. Two FE models of the system under investigation are prepared and discussed. The exact solution is used to verify the developed FE models. The numerical solution demonstrated that the second FE model would be better to simulate the free vibration of the membrane resting on the elastic foundation. Moreover, the second FE model can simulate the vibration of the system with a mass elastic layer. It is important to note that the data presented in this chapter have practical meaning for design engineers.

7. Acknowledgment

MSc Eng Mieczysław Kozłowski and MSc Eng Wojciech Obrocki - WSK PZL Rzeszów S.A. employees - took part in the experimental investigations.

8. References

Allara, M. (2009). A Model for the Characterization of Friction Contacts in Turbine Blades. *Journal of Sound and Vibration,* Vol. 320, No. 3, pp. 527-544, ISSN 0022-460X

Ansell, A. (2005). The Dynamic Element Method for Analysis of Frame and Cable Type Structures. *Engineering Structures,* Vol. 27, No. 13, pp. 1906-1915, ISSN 0141-0296

Chung, W. & Sotelino, E. (2006). Three Dimensional Finite Element Modelling of Composite Girder Bridges. *Engineering Structures,* Vol. 28, No. 1, pp. 63-71, ISSN 0141-0296

De Silva, C. (2005). *Vibration and Shock Handbook,* Taylor & Francis, ISBN 978-0-8493-1580-0, Boca Raton, USA

Jaffrin, M. (2008). Dynamic Shear - Enhanced Membrane Filtration: a Review of Rotating Disks, Rotating Membranes and Vibrating Systems. *Journal of Membrane Science,* Vol. 324, No. 1, pp. 7-25, ISSN 0376-7388

Friswell, M. & Mottershead, J. (1995). *Finite Element Model Updating in Structural Dynamics,* Kluwer Academic Publishers, ISBN 0-7923-3431-0, Dordrecht, Netherlands

Kaliski, S. (1966). *Vibration and Waves in Solids,* IPPT PAN, Warsaw, Poland (in Polish)

Łączkowski, R. (1974). *Vibration of Thermic Turbine Elements,* WNT, Warsaw, Poland (in Polish)

Markowski, T.; Noga, S. & Rudy, S. (2010). Numerical Model of Aviation Gearbox Test Rig in a Closed Loop Configuration. *Aviation,* Vol. 14, No. 1, pp. 3-11, ISSN 1648-7788

Noga, S. (2008). Free Transverse Vibration analysis of an Annular Membrane. *Vibrations in Physical Systems,* Vol. XXIII, pp. 283-288, ISBN 978-83-89333-35-3

Noga, S. (2010). Free Transverse Vibration Analysis of an Elastically Connected Annular and Circular Double - Membrane Compound System. *Journal of Sound and Vibration,* Vol. 329, No. 9, pp. 1507-1522, ISSN 0022-460X (a)

Noga, S. (2010). Free Vibrations of an Annular Membrane Attached to Winkler Foundation. *Vibrations in Physical Systems,* Vol. XXIV, pp. 295-300, ISBN 978-83-89333-35-3 (b)

Rao, S. (2007). *Vibration of Continuous Systems,* Wiley, ISBN-13: 978-0471771715, Hoboken, USA

Rossit, C.; La Malfa, S. & Laura, P. (1998). Antisymmetric Modes of Vibrations of Composite, Doubly - Connected Membranes. *Journal of Sound and Vibration,* Vol. 217, No. 1, pp. 191-195, ISSN 0022-460X

Rudy, S. & Kowalski, T. (1998). Analysis of Contact Phenomenas and Free Vibration Forms of a Blade of Turbine Engine with a Use of FEM, *Rotary Fluid - Flow Machines: proceedings conference,* pp. 105-113, ISBN 83-7199-059-6, Rzeszów, Poland (in Polish)

Tack, J.; Verkerke, G.; van der Houwen, E.; Mahieu, H. & Schutte, H. (2006). Development of a Double - Membrane Sound Generator for Application in a Voice - Producing Element for Laryngectomized Patients. *Annals of Biomedical Engineering,* Vol. 34, No. 12, pp. 1896-1907, ISSN 0090-6964

Toufine, A.; Barrau, J. & Berthillier M. (1999). Dynamic Study of a Simplified Mechanical System with Presence of Dry Friction. *Journal of Sound and Vibration,* Vol. 225, No. 1, pp. 95-109, ISSN 0022-460X

Sinha, S.; Turner, K. (2011). Natural Frequencies of a Pre - twisted Blade in a Centrifugal Force Field. *Journal of Sound and Vibration,* Vol. 330, No. 11, pp. 2655-2681, ISSN 0022-460X

Živanović, S.; Pavic, A. & Reynolds, P. (2007). Finite Element Modelling and Updating of a Lively Footbridge: The Complete Process. *Journal of Sound and Vibration,* Vol. 301, No. 1-2, pp. 126-145, ISSN 0022-460X

Zembaty, Z.; Kowalski, M. & Pospisil, S. (2006). Dynamic Identification of a Reinforced Concrete Frame in Progressive State of Damage. *Engineering Structures,* Vol. 28, No. 5, pp. 668-681, ISSN 0141-0296

9

Torsional Vibration of Eccentric Building Systems

Ramin Tabatabaei
Civil Engineering Department, Islamic Azad University, Kerman Branch, Islamic Republic of Iran

1. Introduction

The comprehensive studies conducted by a number of researchers in the past few decades and investigations of the effects of past earthquakes have shown that in buildings with non-coincident the center of mass (CM) and the center of rigidity (CR), significant coupling may occur between the translational and the torsional displacements of the floor diaphragms even when the earthquake induces uniform rigid base translations (Kuo, 1974; Chandler & Hutchinson, 1986; Cruz & Chopra, 1986; Hejal & Chopra, 1989).

In investigating the seismic torsional response of structures to earthquakes, it is customary to assume that each point of the foundation of the structure is excited simultaneously. Under this assumption, if centers of mass and rigidity of the floor diaphragms lie along the same vertical axis, a horizontal component of ground shaking will induce only lateral or translational components of motion. On the other hand, if the centers of mass and rigidity do not coincide, a horizontal component of excitation will generally induce both lateral components of motion and a rotational component about a vertical axis. Structures for which the centers of mass and rigidity do not coincide will be referred to herein as eccentric structures. Torsional actions may also be induced in symmetric structures due to the fact that, even under a purely translational component of ground excitation, all points of the base of the structure are not excited simultaneously because of the finite speed of propagation of the ground excitation, (Kuo, 1974).

This seismic torsional response leads to increased displacement at the extremes of the torsionally asymmetric building systems and may cause suffering in the lateral load-resisting elements located at the edges, particularly in the systems that are torsionally flexible. More importantly, the seismic response of the systems, especially in the torsionally flexible structure is qualitatively different from that obtained in the case of static loading at the center of mass. To account for the possible amplification in torsion produced by seismic response and accidental torsion in the elastic range, the equivalent static eccentricities of seismic forces are usually defined by building codes with simple expressions of the static eccentricity. The equivalent static eccentricities of seismic forces are proposed by researchers, (Dempsey & Irvine, 1979, Tso & Dempsey, 1980 and De la Llera & Chopra, 1994). A clear and comprehensive study of the equivalent static eccentricities that are presented by Anastassiadis et al., (1998), included a set of formulas for a one-storey scheme, allow the evaluation of the exact additional eccentricities necessary to be obtained by means of static analysis the maximum displacements at both sides of the deck, or the maximum deck rotation, given by modal analysis. A procedure to extend the static torsional provisions of code to asymmetrical multi-storey buildings is presented by Moghadam and Tso, (2000). They have developed a refined method for determination of CM eccentricity and torsional radius for multi-storey buildings. However, the inelastic torsional response is less easily predictable, because the location of the center of rigidity on each floor cannot be determined readily and the equivalent static eccentricity varies storey by storey at each nonlinear static analysis step. The simultaneous presence of two orthogonal seismic components or the contemporary eccentricity in two orthogonal directions may have some importance, mainly in the inelastic range, (Fajfar et al., 2005). Consequently, the static analysis with the equivalent static eccentricities can be effective only if used in the elastic range. This can only be achieved, the location of the static eccentricity is necessary to change in each step of the nonlinear static procedure. It may be needed for the development of simplified nonlinear assessment methods based on pushover analysis.

Fig. 1. Damage to buildings subjected to strong earthquakes, (9-11 Research Book, 2006)

However, the seismic torsional response of asymmetric buildings in the inelastic range is very complex. The inelastic response of eccentric systems only has been investigated in an exploratory manner, and, on the whole, it has not been possible to derive any general conclusions from the data that were obtained. No work appears to have been reported concerning the torsional effects induced in symmetric structures deforming into the inelastic range (Tanabashi, 1960; Koh et al., 1969; Fajfar et al., 2005).

Torsional motion is produced by the eccentricity existing between the center of mass and the center of rigidity. Some of the situations that can give rise to this situation in the building plan are:

- Positioning the stiff elements asymmetrically with respect to the center of gravity of the floor.
- The placement of large masses asymmetrically with respect to stiffness.

- A combination of the two situations described above.

Consequently, torsional-translational motion has been the cause of major damage to buildings vibrated by strong earthquakes, ranging from visible distortion of the structure to structural collapse (see Fig. 1). The purpose of this chapter is to investigate the torsional vibration of both symmetric and eccentric one-storey building systems subjected to the ground excitation.

Fig. 2. Mexico City building failure associated with the torsional-translation motion, (Earthquake Engineering ANNEXES, 2007)

2. Classification of vibration

Vibration can be classified in several ways. Some of the important classifications are as follows: Free and forced vibration: If a system, after an internal disturbance, is left to vibrate on its own, the ensuing vibration is known as free vibration. No external force acts on the system. The oscillation of the simple pendulum is an example of free vibration.

If a system is subjected to an external force (often, a dynamic force), the resulting vibration is known as forced vibration. The oscillation that arises in buildings such as earthquake is an example of forced vibration.

A building, for which the centers of mass and rigidity do not coincide, (eccentric building) will experience a coupled torsional-translational motion even when it is excited by a purely translational motion of the ground. The torsional component of response may contribute significantly to the overall response of the building, particularly when the uncoupled torsional and translational frequencies of the system are close to each other (see Fig. 2).

Failures of such structures as buildings and bridges have been associated with the torsional-translational motion.

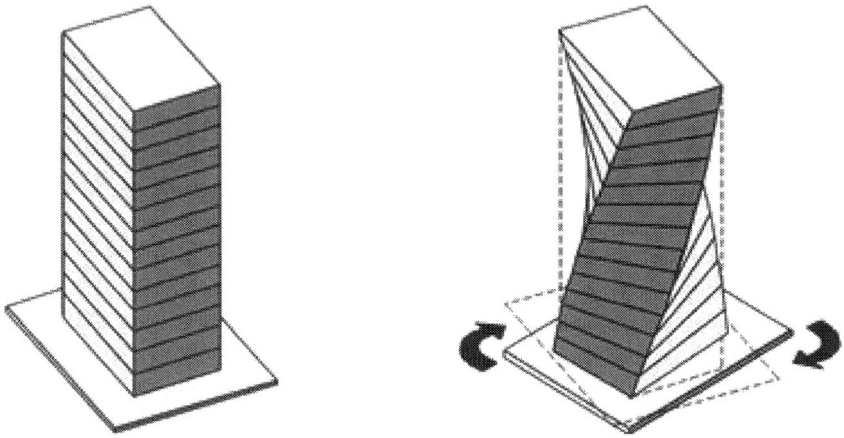

Fig. 3. Torsional vibration mode shape

2.1 Free vibration analysis

One of the most important parameters associated with engineering vibration is the natural frequency. Each structure has its own natural frequency for a series of different mode shapes such as translational and torsional modes which control its dynamic behaviour (see Fig. 3). This will cause the structures to be subjected to series structural vibrations, when they are located in environments where earthquakes or high winds exist. These vibrations may lead to serious structural damage and potential structural failure.

In buildings, both translational and torsional vibration modes arise, even if, little eccentricity in the transverse direction during earthquakes. The in-plane floor vibration mode such as arch-shaped floor vibration mode also arises during earthquakes. However these observational data are not enough at present. The causes of the torsional-translational vibration are thought as follows:

1. Input motion to the foundation has a possibility to contain the torsional component, which is the cause of the torsional vibration.
2. The torsional coupling, due to the eccentricity in both directions, is also a cause of the torsional vibration. It arises surely when the eccentricity in the transverse direction is large. However, even if the eccentricity is small, it is well-known that the strong torsional coupling also arises when the natural frequencies of the translational mode and the torsional mode approach closely to each other.
3. The eccentricity in the transverse direction is small in general, since sufficient attention is usually paid on the eccentricity to prevent the torsional vibration in the structural planning. On the other hand, the eccentricity in the longitudinal direction results often from necessity of architectural planning and/or from insufficiency of attention on the eccentricity in the structural planning, but it is also small as a necessity from the configuration of the floor plan.

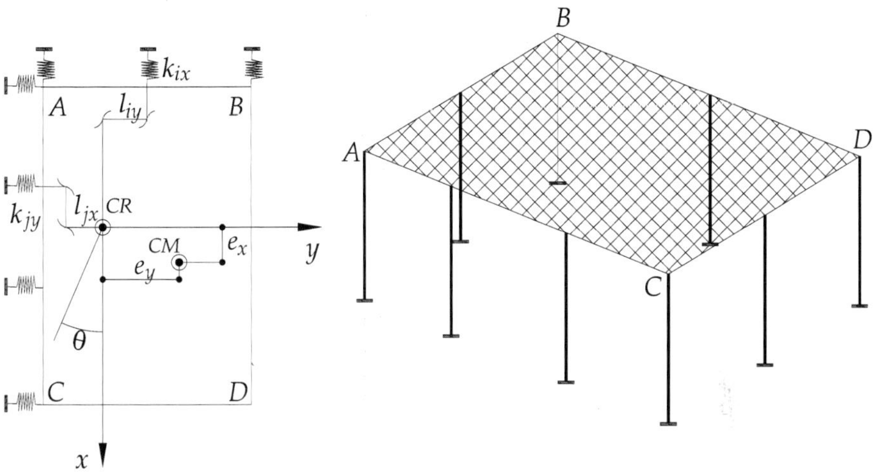

Fig. 4. Model of a one-storey system with double eccentricities

2.1.1 One-storey system with double eccentricities

The estimation of torsional-translational response of simplified procedure subjected to a strong ground motion, is a key issue for the rational seismic design of new buildings and the seismic evaluation of exacting buildings. This section is a vibration-based analysis of the simple one-storey model with double eccentricities, and it would be a promising candidate as long as buildings oscillate predominantly in the two lateral directions (Tabatabaei and Saffari, 2010).

2.1.2 Basic parameters of the model

The one-storey system, considered in this section, may be modeled as shown in Fig. 4. The center of rigidity (CR) is the point in the plan of the rigid floor diaphragm through which a lateral force must be applied in order that it may cause translational displacement without torsional rotation. When a system is subjected to forces, which will cause pure rotation, the rotation takes place around the center of rigidity, which remains fixed. The location of the center of rigidity can be determined from elementary principles of mechanic.

The horizontal rigid floor diaphragm is constrained in the two lateral directions by resisting elements (columns). Let k_{ix} and k_{jy} be the lateral stiffness of the i-th and j-th resisting element in x-direction and y-direction, respectively. The origin of the coordinates is taken at the center of rigidity (CR). A system for which the eccentricities, e_x and e_y are both different from zero, has three degrees of freedom. Its configuration is specified by translations x and y and rotation, θ. The positive directions of these displacements are indicated on the figure.

Applying the geometric relationships between the centers of mass and rigidity, the equations of motion of undamped free vibration of the system may be written as follows

$$m(\ddot{x}+e_y\ddot{\theta})+K_x x=0 \tag{1a}$$

$$m(\ddot{y}-e_x\ddot{\theta})+K_y y=0 \tag{1b}$$

$$I_m\ddot{\theta}+K_\theta\theta-me_x(\ddot{y}-e_x\ddot{\theta})+me_y(\ddot{x}+e_y\ddot{\theta})=0 \tag{1c}$$

where

$K_x=\sum_{i=1}^{n}k_{ix}$: total translational stiffness in the x-direction (n = number of columns in x-dir),

$K_y=\sum_{j=1}^{m}k_{jy}$: total translational stiffness in the y-direction (m = number of columns in y-dir),

$K_\theta=\sum_{i=1}^{n}k_{ix}l_{iy}^2+\sum_{j=1}^{m}k_{jy}l_{jx}^2$: total rotational stiffness, m: total mass, I_m: the mass moment of inertia of the system around the center of mass (CM), and l_{iy} and l_{jx}, be the distances of the i-th and j-th resisting element from the center of rigidity along the x and y axes, as shown in Fig. 4.

For free vibration analysis, the solution of Eqs. (1) may be taken in the form

$$x=X\sin(\omega t) \tag{2a}$$

$$y=Y\sin(\omega t) \tag{2b}$$

$$\theta=\Theta\sin(\omega t) \tag{2c}$$

where X, Y and Θ are the displacements amplitudes in x, y and θ directions, respectively. The value of ω is referred to the circular natural frequency. Substitution of Eqs. (2) into Eqs. (1) given in

$$(-m\omega^2+K_x)X-m\omega^2e_y\Theta=0 \tag{3a}$$

$$(-m\omega^2+K_y)Y+m\omega^2e_x\Theta=0 \tag{3b}$$

$$(-(I_m+me_x^2+me_y^2)\omega^2+K_\theta)\Theta-m\omega^2e_yX+m\omega^2e_xY=0 \tag{3c}$$

Eqs. (3) have a nontrivial solution only if the determinate of the coefficients of X, Y and Θ are equal to zero. This condition yields the characteristic equation of describing such a system may be taken in the form

$$\omega^6-\left[\frac{K_x+K_y}{m}+\frac{K_\theta}{I_m}+\frac{K_xe_y^2}{I_m}+\frac{K_ye_x^2}{I_m}\right]\omega^4+\left[\frac{K_xK_y}{m^2}+\frac{K_\theta(K_x+K_y)}{mI_m}+\frac{K_xK_ye_y^2}{mI_m}+\frac{K_xK_ye_x^2}{mI_m}\right]\omega^2 -\frac{K_xK_yK_\theta}{m^2I_m}=0 \tag{4}$$

where, e_x is the static eccentricity (eccentricity between mass and rigidity centers) in the x-direction and e_y is the static eccentricity in the y-direction. Now letting the following expressions,

$$\omega_x^2 = \frac{K_x}{m} \quad \omega_y^2 = \frac{K_y}{m} \quad \omega_\theta^2 = \frac{K_\theta}{I_m} \tag{5a}$$

$$\varepsilon_x = \frac{e_x}{r_m} \quad \varepsilon_y = \frac{e_y}{r_m} \quad c = \sqrt{1 + \varepsilon_x^2 + \varepsilon_y^2} \tag{5b}$$

$$\varepsilon = \frac{e}{r_m} \quad e = \sqrt{e_x^2 + e_y^2} \tag{5c}$$

and making use of the relation $I_m = m \cdot r_m^2$; , where r_m is the radius gyration of mass, Eq. (4) may be written in the following dimensionless form:

$$\left(\frac{\omega}{\omega_x}\right)^6 - \left[1 + \varepsilon_x^2 + \left(\frac{\omega_y}{\omega_x}\right)^2\left(1 + \varepsilon_y^2\right) + \left(\frac{\omega_\theta}{\omega_x}\right)^2\right]\left(\frac{\omega}{\omega_x}\right)^4 + \left[\left(c\frac{\omega_y}{\omega_x}\right)^2 + \left(\frac{\omega_\theta}{\omega_x}\right)^2\left(1 + \left(\frac{\omega_y}{\omega_x}\right)^2\right)\right]\left(\frac{\omega}{\omega_x}\right)^2 - \left(\frac{\omega_y}{\omega_x}\right)^2\left(\frac{\omega_\theta}{\omega_x}\right)^2 = 0 \tag{6}$$

where the values of ω_x and ω_y are referred to the uncoupled circular natural frequencies of the system in x and y-directions, respectively. The value of ω_θ will be referred as the uncoupled circular natural frequency of torsional vibration. The n - th squares of the coupled natural frequency ω_n are defined by three roots of the characteristic equation defined in Eq. (6). Associated with each natural frequency, there is a natural mode shape vector $\{\varphi_n\} = \{\varphi_{xn}, \varphi_{yn}, \varphi_{\theta n}\}^T$ of the one-storey asymmetric building models that can be obtained with assuming, $\{\varphi_{xn}\} = 1$, and two components as follows,

$$\{\varphi_{yn}\} = \frac{\varepsilon_y}{\varepsilon_x}\left[\frac{1 - \left(\omega_n / \omega_x\right)^2}{\left(\omega_n / \omega_x\right)^2 - \left(\omega_y / \omega_x\right)^2}\right] \tag{7a}$$

$$\{\varphi_{\theta n}\} = \frac{c}{\varepsilon_x}\left[\frac{1}{\left(\omega_n / \omega_x\right)^2} - 1\right] / r_{cr} \tag{7b}$$

where n varies from 1 to 3 and $r_{cr} = \sqrt{r_m^2 + e_x^2 + e_y^2}$, (Kuo, 1974).

As a matter of fact, the numerical results have been evaluated over a wide range of the frequency ratio ω_θ / ω_x for several different values of eccentricity parameter ε_y. A value of

$e_x/e_y = 1$ which corresponds to systems with double eccentricities along the x-axis and y-axis is considered. In the latter case, two values of ω_y/ω_x are considered. The coupled natural frequencies are summarized in Figs. 5 and 6 are also applicable to the system considered in this section for any given longitudinal distribution of motions.

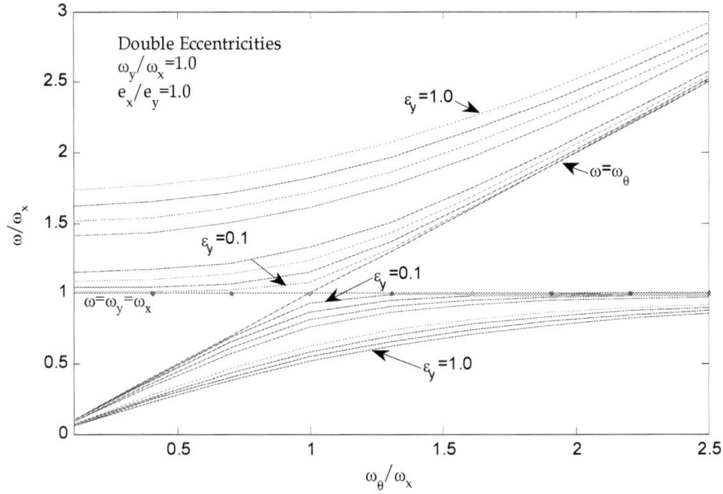

Fig. 5. The coupled natural frequency ratio for varying eccentricity parameter, ε_y of Double eccentricities system and $\omega_y/\omega_x = 1.0$

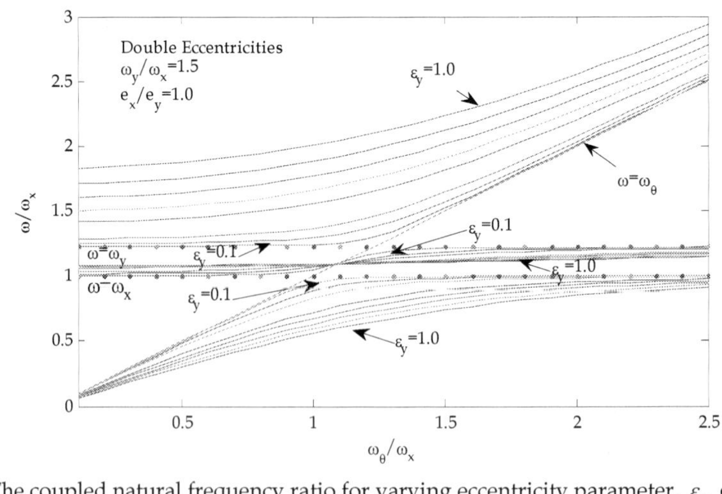

Fig. 6. The coupled natural frequency ratio for varying eccentricity parameter, ε_y of Double eccentricities system and $\omega_y/\omega_x = 1.5$

In these Figures, the uncoupled natural frequencies of the systems are represented by the straight lines corresponding to $\varepsilon_y = 0$. For the systems with double eccentricity considered in Fig. 5, these are defined by the diagonal line and the two horizontal lines. The diagonal line represents the uncoupled torsional frequency, and the horizontal lines the two uncoupled translational frequencies. As would be expected, the lower natural frequency of the coupled system is lower than either of the frequencies of the uncoupled system. Similarly, the upper natural frequency of the coupled system is higher than the upper natural frequency of the uncoupled system. The general trends of the curves for the coupled systems are typical of those obtained for other combinations of the parameters as well.

The curve for the lowest frequency always starts from the origin whereas the curve for the highest frequency starts from a value higher than the uncoupled translational frequencies of the system, depending on the value of the eccentricity. Both curves increase with the higher value of ω_θ/ω_x. For 1arge value of ω_θ/ω_x the lowest frequency approaches the value of ω_x and the highest frequency approaches the value of ω_θ. The maximum coupling effect on frequencies occurs when the value of ω_θ/ω_x is equal to unity, (Kuo, 1974).

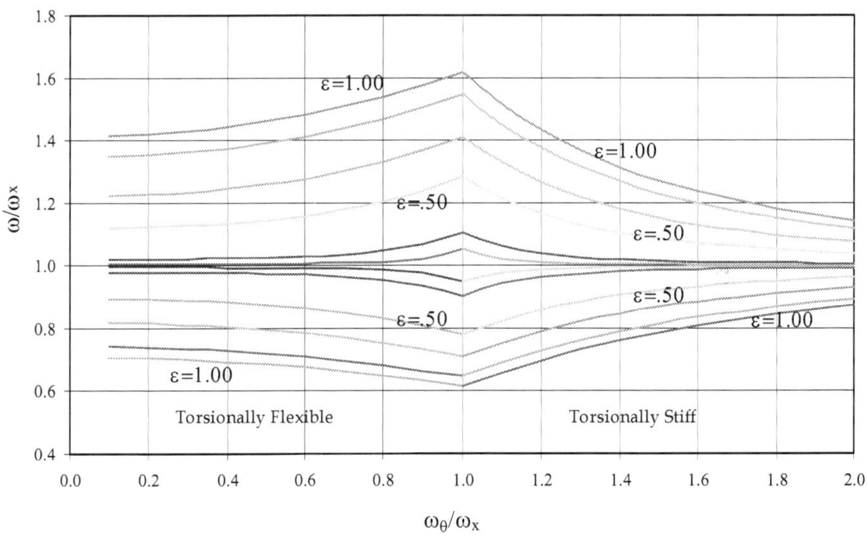

Fig. 7. Relationship between Coupled and Uncoupled Natural Frequencies

It is interesting to note that the coupled dynamic properties depend only on the four dimension 1ess parameters ε_x, ε_y, ω_θ/ω_x and ω_y/ω_x. Fig. 7 shows the relationship between the coupled and uncoupled natural frequencies, in one way torsionally coupled systems (with $\varepsilon_x = 0$), for different values of ε.

If ρ_n represents the distance positive to the left from the center of mass to the instantaneous center of rotation of the system for the modes under consideration, it can be shown that (see Fig. 8).

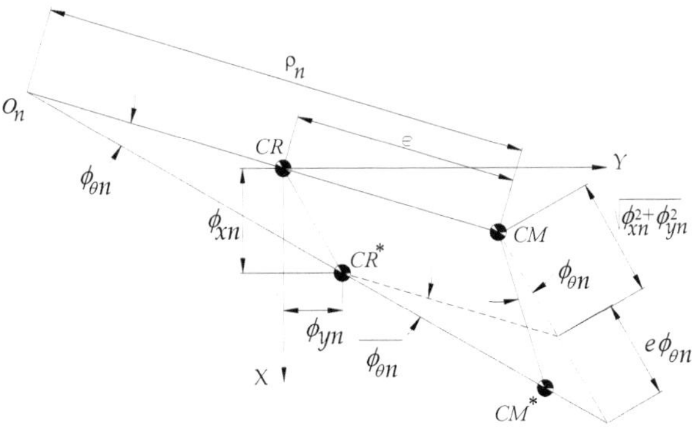

Fig. 8. CR^* and CM^* denote the new locations of the centers of rigidity and mass at any time instant, respectively (Tabatabaei and Saffari, 2010).

$$\frac{\rho_n}{e} = 1 + \frac{\varphi_{xn}}{e\varphi_{\theta n}}\sqrt{1 + \left(\varphi_{yn}\big/\varphi_{xn}\right)^2} \tag{8}$$

The ratio of $\rho_n/e = 1$ indicates that the center of rotation is at the center of rigidity, whereas the value of $\rho_n/e = 0$ indicates that the center of rotation is at the center of mass. By making use of Eq. (7), Eq. (8) may also be related to the frequency values.

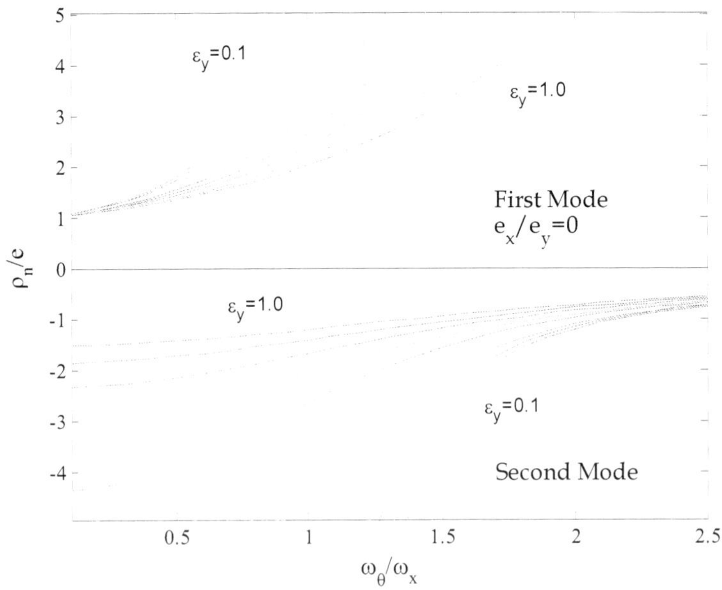

Fig. 9. Location of the center of the rotation normalized with the respect to eccentricity

For the one-storey system with eccentricity considered in Fig. 4, the locus of the associated center of rotation is plotted in Fig. 9. It should be recalled that a value of $\rho_n/e = 1$ in the latter figure indicates that the center of rotation is at the center of rigidity of the system. Note that as the value of ω_θ/ω_x increases, the center of rotation shifts away from the center of rigidity for the first mode and approaches the center of mass for the higher mode for all values of eccentricity.

2.1.3 One-storey system with single eccentricity

In the particularly case of $e_x = 0$, system with single eccentricity and second equation in Eq. (1) becomes independent of the others. The motion of the system in this case is coupled only in the x and θ directions. The following frequency equation is obtained from Eq. (6) by taking $\varepsilon_x = 0$ and factoring out the term $(\omega^2 - \omega_x^2)/\omega_y^2$, which obviously defines the uncoupled natural frequency of the system in the y direction:

$$\left(\frac{\omega}{\omega_x}\right)^4 - \left[1 + \varepsilon_y^2 + \left(\frac{\omega_\theta}{\omega_x}\right)^2\right]\left(\frac{\omega}{\omega_x}\right)^2 + \left(\frac{\omega_\theta}{\omega_x}\right)^2 = 0 \qquad (9)$$

Numerical data have been evaluated over a wide range of the frequency ratio ω_θ/ω_x for several different values of eccentricity parameter ε_y. Two values of $\varepsilon_x/\varepsilon_y$ are considered: a value of $\varepsilon_x/\varepsilon_y = 0$ which corresponds to systems with an eccentricity along the y-axis and a value of $\varepsilon_x/\varepsilon_y = 1$. In the latter case, two values of ω_y/ω_x are considered. The coupled natural frequencies are summarized in Fig 10.

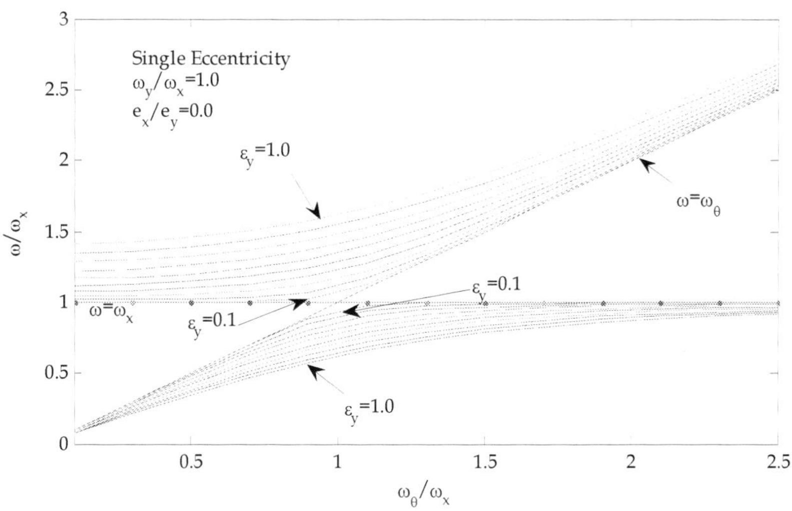

Fig. 10. The coupled natural frequency ratio for several different values of eccentricity parameter, ε_y of single eccentricity system

As a result, Figs. 5 and 10, which refer to systems with double and single eccentricities, respectively, are very similar in form. This is due to the fact that, since $\omega_y/\omega_x = 1$, the frequencies of the two uncoupled translational modes are represented by the same horizontal line, $\omega = \omega_x = \omega_y$ and this frequency value is also equal to the second natural frequency of the coupled system. As a matter of fact, it can be shown that the curves in Fig. 10 may be obtained from those in Fig. 5 by interpreting ε_y to be equal to $\varepsilon_y\sqrt{1+\left(\varepsilon_x/\varepsilon_y\right)^2}$, (Kuo, 1974).

2.1.4 Classification of torsionl behaviour

To obtain the knowledge on the torsional response of buildings, a key elastic parameter is the ratio of the two uncoupled frequencies, Ω given by

$$\Omega = \frac{\omega_\theta}{\omega_x} \tag{10}$$

where ω_x and ω_θ, are an uncoupled translational and torsional frequencies, respectively.
If Ω is greater than 1 the response is mainly translational and structure behaviour is defined "torsionally stiff"; conversely if Ω is lower than 1 the response is affected to a large degree by torsional behaviour and, then, the structure is defined as "torsionally flexible". A clear and comprehensive study of this subject is presented in Ref., (Anastassiadis et al., 1998). In the case of a torsionally stiff structure, a single translation mode controls the displacement in on direction. Thus the typical dynamic behaviour of such structure is qualitatively similar to the response obtained using static analysis, i.e. the displacements increase at the flexible edge (side is outlying to the center of rigidity), and decrease at the stiff edge (side is closed to the center of rigidity). The seismic response of torsionally flexible structure is qualitatively different from that obtained in the case of static loading at the center of mass. The main reason is that the displacement envelope of the deck depends on both the translation and the torsional modes.

2.2 Forced vibration analysis

Civil engineering structures are always designed to carry their own dead weight, superimposed loads and environmental loads such as wind or earthquake. These loads are usually treated as maximum loads not varying with time and hence as static loads. In some cases, the applied load involves not only static components, but also contains a component varying with time which is a dynamic load. In the past, the effects of dynamic loading have often been evaluated by use of an equivalent static load, or by an impact factor, or by a modification of the factor of safety.
Many developments have been carried out in order to try to quantify the effects produced by dynamic loading. Examples of structures, where it is particularly important to consider dynamic loading effects, are the construction of tall buildings, long bridges under wind-loading conditions, and buildings in earthquake zones, etc.
Typical situations, where it is necessary to consider more precisely, the response produced by dynamic loading are vibrations due to earthquakes. So it is very important to study the dynamic nature of structures.

Dynamic characteristics of damaged and undamaged buildings are, as a rule, different. This difference is caused by a change in stiffness and can be used for the detection of damage and for the determination of some damage parameters such as crack magnitude and location. In this connection, the use of vibration methods of damage diagnostics is promising. These methods are based on the relationships between the vibration characteristics (natural frequencies and mode shapes) or peculiarities of vibration system behaviour (for example, drift of building edges, the amplitudes of base shear, the resonance frequencies, etc.) and damage parameters.

Depending on the assumptions adopted, the type of analysis used the kind of the loading or excitation and the overall building characteristics, a variety of different approaches have been reported in the references to one-storey building systems.

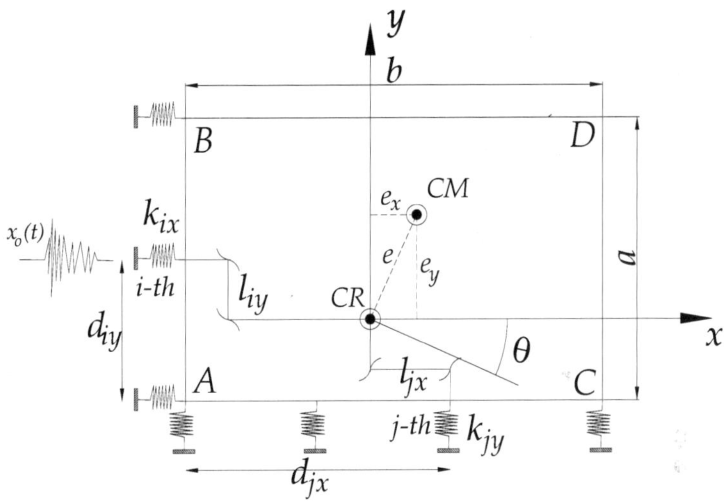

Fig. 11. Model of a one-storey system with double eccentricities subjected to ground motion

2.2.1 One-storey system with double eccentricities

In this section, the dynamic response of one-storey system to the horizontal components of ground motion is considered. At any instant of time, the ground motion is assumed to be the same at all points on the foundation. Also, the ground motion is considered to be plane shear waves propagating horizontally with a constant velocity and without change in shape. Now, equations of motion are developed for system as shown in Fig. 11 subjected to a horizontally propagating ground excitation. Let the earthquake ground motion be defined by accelerations along the two axes. Therefore, a force applied along either of the two principal axes of rigidity will cause displacement in the same direction. The principal axes of rigidity are orthogonal and pass through the center of rigidity. For building plans of the type shown in Fig. 11, where principal axes of rigidity of individual column sections are all parallel to one another, the principal axes of rigidity of the complete building are parallel to those of the individual elements. Within the range of linear behaviour, the equations of motion of the system, written about the center of rigidity of the system, are as follows

$$m\left(\ddot{x}+e_y\ddot{\theta}\right)+C_x\dot{x}+K_x x-\sum_{i=1}^{n}c_{ix}\left(\dot{x}_g\right)_i-\sum_{i=1}^{n}k_{ix}\left(x_g\right)_i=0 \tag{11a}$$

$$m\left(\ddot{y}-e_x\ddot{\theta}\right)+C_y\dot{y}+K_y y-\sum_{j=1}^{m}c_{jy}\left(\dot{y}_g\right)_j-\sum_{j=1}^{m}k_{jy}\left(y_g\right)_j=0 \tag{11b}$$

$$\begin{aligned}&I_m\ddot{\theta}+C_\theta\dot{\theta}+R\theta+me_y\left(\ddot{x}+e_y\ddot{\theta}\right)-me_x\left(\ddot{y}-e_x\ddot{\theta}\right)-\sum_{i=1}^{n}c_{ix}\left(\dot{x}_g\right)_i l_{iy}+\sum_{j=1}^{m}c_{jy}\left(\dot{y}_g\right)_j l_{jx}\\&-\sum_{i=1}^{n}k_{ix}\left(x_g\right)_i l_{iy}+\sum_{j=1}^{m}k_{jy}\left(y_g\right)_j l_{jx}=0\end{aligned} \tag{11c}$$

where, x and y are horizontal displacements of the center of rigidity of the deck, relative to the ground, along the principal axes of rigidity of the model, x and y, and θ are the rotation of the deck about the vertical axis; $\left(x_g\right)_i$ and $\left(y_g\right)_j$ are the ground displacements at the i - th and j - th column supports in the x and y directions, respectively; C_x, C_y and C_θ are the total damping coefficients associated with the motions in the x, y and θ directions, respectively; c_{ix} and c_{jy} are the individual damping coefficients of the i - th and j - th columns in the x and y directions, respectively; k_{ix} and k_{jy} are the individual lateral stiffness of the i - th and j - th columns in the x and y directions, respectively; and l_{ix} and l_{jy} are the normal distances measured from the center of rigidity to the i - th and j - th columns in the y and x directions, respectively.

It is important to note that time is measured from the instant the ground motion reaches the first support on the left of the diagram shown in Fig. 11 and that the amplitude of the ground motion at any support is a function of both time and distance from the first, i.e.,

$$\left(\dot{x}_g\right)_i=x_g\left(t-\frac{d_{iy}}{v}\right) \quad for \quad t\geq\frac{d_{iy}}{v} \tag{12a}$$

$$=0 \qquad for \quad t<\frac{d_{iy}}{v}$$

$$\left(\dot{y}_g\right)_j=y_g\left(t-\frac{d_{jx}}{v}\right) \quad for \quad t\geq\frac{d_{jx}}{v} \tag{12b}$$

$$=0 \qquad for \quad t<\frac{d_{jx}}{v}$$

where v is the shear wave velocity of the ground motion; and d_{iy} and d_{jx} are the distances from the first support to the i - th and j - th column support in the x and y directions, respectively.

Since it is customary to specify the ground motion in terms of its acceleration, Eqs. (11) will be rewritten by making a proper coordinate transformation. Assuming that the damping coefficients are proportional to the corresponding stiffness coefficients, and letting

$$U_x = x - \frac{\sum_{i=1}^{n} k_{ix}\left(x_g\right)_i}{K_x} \tag{13a}$$

$$U_y = y - \frac{\sum_{j=1}^{m} k_{jy}\left(y_g\right)_j}{K_y} \tag{13b}$$

$$U_\theta = \theta - \frac{1}{K_\theta}\left[\sum_{i=1}^{n} k_{ix}\left(x_g\right)_i l_{iy} - \sum_{j=1}^{m} k_{jy}\left(y_g\right)_j l_{jx}\right] \tag{13c}$$

where n and m are the total numbers of the column support in the x and y directions, respectively. Also, physically, U_x, U_y and U_θ are the displacements of the system relative to the instantaneous value of the average amplitude of the ground motion experienced by the base.

Substituting Eqs. (13) in Eqs. (11) and writing those equations, after introducing damping, in a convenient form yields

$$\left(\ddot{U}_x + e_y\ddot{U}_\theta\right) + 2\beta_x\omega_x\dot{U}_x + \omega_x^2 U_x = -\frac{\sum_{i=1}^{n} k_{ix}\left(\ddot{x}_g\right)_i}{K_x} - \frac{e_y}{K_\theta}\left[\sum_{i=1}^{n} k_{ix}\left(\ddot{x}_g\right)_i l_{iy} - \sum_{j=1}^{m} k_{jy}\left(\ddot{y}_g\right)_j l_{jx}\right] \tag{14a}$$

$$\left(\ddot{U}_y - e_x\ddot{U}_\theta\right) + 2\beta_y\omega_y\dot{U}_y + \omega_y^2 U_y = -\frac{\sum_{j=1}^{m} k_{jy}\left(\ddot{y}_g\right)_j}{K_y} + \frac{e_x}{K_\theta}\left[\sum_{i=1}^{n} k_{ix}\left(\ddot{x}_g\right)_i l_{iy} - \sum_{j=1}^{m} k_{jy}\left(\ddot{y}_g\right)_j l_{jx}\right] \tag{14b}$$

$$\begin{aligned} \ddot{U}_\theta + \frac{2}{c^2}\beta_\theta\omega_\theta\dot{U}_\theta + \frac{1}{c^2}\omega_\theta^2 U_\theta + \frac{me_y}{I_{cr}}\ddot{U}_x - \frac{me_x}{I_{cr}}\ddot{U}_y = \frac{me_x}{I_{cr}}\frac{\sum_{j=1}^{m} k_{jy}(\ddot{y}_g)_j}{K_y} \\ -\frac{me_y}{I_{cr}}\frac{\sum_{i=1}^{n} k_{ix}(\ddot{x}_o)_i}{K_x} - \frac{1}{K_\theta}\left[\sum_{i=1}^{n} k_{ix}\left(\ddot{x}_g\right)_i l_{iy} - \sum_{j=1}^{m} k_{jy}\left(\ddot{y}_g\right)_j l_{jx}\right] \end{aligned} \tag{14c}$$

where β_x, β_y and β_θ are the damping coefficients in fraction of the critical damping for vibrations in the x, y and θ directions, respectively.

Eqs. (14) has been expressed in the most general form. It is applicable to both symmetric and eccentric systems subjected to a ground disturbance having a finite speed of propagation. It may be reduced to the equations derived from a conventional analysis in which the speed of propagation of the ground motion is assumed to be infinite.

2.2.2 Time history response

The modal equations (14) can be solved numerically using the modal superposition method or by direct numerical integration. The modal superposition method, which involves the use of the characteristic values and functions of the system, uncouples the equations of motion so that each of the uncoupled equations may be integrated independently. Since it is based on the assumption that the structure behaves linearly, this method is applicable only to the elastic range of response. These issues have been further discussed in Ref., (Kuo, 1974).

The method of direct numerical integration, which integrates the equations of motion in their original form, may be applied to both the elastic and inelastic ranges of response. For the inelastic analysis, the material properties of the system are assumed to be linear within a small time interval, and they are modified at the end of each integration step when needed.

The modal superposition method is used to evaluate the elastic response of both symmetric and eccentric systems. Before solving Eqs. (14), it is desirable to rewrite it in matrix form. For consistency in units, the third equation is multiplied by r_{cr} to give the following set:

$$\begin{bmatrix} 1 & 0 & e_y/r_{cr} \\ 0 & 1 & -e_x/r_{cr} \\ e_y/r_{cr} & -e_x/r_{cr} & 1 \end{bmatrix} \begin{Bmatrix} \ddot{U}_x \\ \ddot{U}_y \\ r_{cr}\ddot{U}_\theta \end{Bmatrix} + \begin{bmatrix} 2\beta_x\omega_x & 0 & 0 \\ 0 & 2\beta_x\omega_x & 0 \\ 0 & 0 & 2\beta_\theta\omega_\theta/c^2 \end{bmatrix} \begin{Bmatrix} \dot{U}_x \\ \dot{U}_y \\ r_{cr}\dot{U}_\theta \end{Bmatrix} + \begin{bmatrix} \omega_x^2 & 0 & 0 \\ 0 & \omega_y^2 & 0 \\ 0 & 0 & \omega_\theta^2/c^2 \end{bmatrix} \begin{Bmatrix} U_x \\ U_y \\ r_{cr}U_\theta \end{Bmatrix}$$
$$= \begin{Bmatrix} -\sum_{i=1}^{n} k_{ix}^{*}\left(\ddot{x}_g\right)_i - \frac{e_y}{b}\ddot{T}_\theta \\ -\sum_{j=1}^{m} k_{jy}^{*}\left(\ddot{y}_g\right)_j - \frac{e_x}{b}\ddot{T}_\theta \\ \frac{e_x}{r_{cr}}\sum_{j=1}^{m} k_{jy}^{*}\left(\ddot{y}_g\right)_j - \frac{e_y}{r_{cr}}\sum_{i=1}^{n} k_{ix}^{*}\left(\ddot{x}_g\right)_i - \frac{r_{cr}}{b}\ddot{T}_\theta \end{Bmatrix} \tag{15}$$

where the uncoupled circular natural frequencies ω_x, ω_y and ω_θ are defined by Eq. (5a), and c is defined by Eq. (5b). Other symbols are defined as follows:

$$k_{ix}^{*} = \frac{k_{ix}}{K_x}, \quad k_{jy}^{*} = \frac{k_{jy}}{K_y} \tag{16a}$$

$$\ddot{T}_\theta = \frac{\sum_{i=1}^{n} k_{ix}^{*}\left(\ddot{x}_g\right)_i \eta_i - \frac{a}{b}\frac{K_y}{K_x}\sum_{j=1}^{m} k_{jy}^{*}\left(\ddot{y}_g\right)_j \xi_j}{\sum_{i=1}^{n} k_{ix}^{*}\eta_i^2 + \left(\frac{a}{b}\right)^2 \frac{K_y}{K_x}\sum_{j=1}^{m} k_{jy}^{*}\xi_j^2} \tag{16b}$$

$$\eta_i = \frac{l_{iy}}{b}, \quad \xi_j = \frac{l_{jx}}{a} \tag{16c}$$

Symbolically, Eq. (15) may also be written as

$$[m]\{\ddot{U}\}+[C]\{\dot{U}\}+[K]\{U\}=\{f(t)\} \tag{17}$$

Now, let $[\varphi]$ be the modal matrix, formed of the three natural modes of the system, and let

$$\{U\}=[\varphi]\{Z\} \tag{18}$$

Substituting Eq. (18) into Eq. (17), multiplying the resulting equation by $[\varphi]^T$, and making use of the orthogonality of the natural modes, Eq. (17) leads to the independent modal equations as follows

$$\ddot{Z}_n+2\beta_n\omega_n\dot{Z}_n+\omega_n^2 Z_n=\frac{\{\varphi_n\}^T\{f(t)\}}{\{\varphi_n\}^T[m]\{\varphi_n\}} \tag{19}$$

where n = 1,2,3; ω_n is the n-th coupled natural frequency of the system; β_n is the damping coefficient associated with the nth natural mode; and $\{\varphi_n\}$ is the n-th natural mode.
It should be recalled that the damping coefficients are assumed to be proportional to the stiffness coefficients. For specified β_n and $f(t)$, each of the three equations may then be solved independently by a step-by-step numerical integration procedure.
There are several different methods available for integrating numerically equations of the form of Eq. (19). One of the procedures is the linear acceleration method, which is simple but sufficiently accurate for all practical purposes if the integration increment is chosen suitably.
In this method, the acceleration is assumed to vary linearly with time, t. Let $Z_n(t)$, $\dot{Z}_n(t)$ and $\ddot{Z}_n(t)$ denote, respectively, the value of Z_n and of its first two derivatives at any time t, and let Δt be the same increment between t and $t+\Delta t$. The acceleration Z_n at time $t+\Delta t$ may then be expressed in terms of all previous values at time t as follows:

$$\ddot{Z}_n(t+\Delta t)=\frac{1}{m_{eq}}\left[F_n(t+\Delta t)-2\beta_n\omega_n\dot{q}_n-\omega_n^2 q_n\right] \tag{20}$$

where

$$F_n(t+\Delta t)=\frac{\{\varphi_n\}^T\{f(t+\Delta t)\}}{\{\varphi_n\}^T[m]\{\varphi_n\}} \tag{21a}$$

$$\dot{q}_n=\dot{Z}_n(t)+\frac{\Delta t}{2}\ddot{Z}_n(t) \tag{21b}$$

$$q_n=Z_n(t)+\Delta t\dot{Z}_n(t)+\frac{\Delta t^2}{3}\ddot{Z}_n(t) \tag{21c}$$

$$m_{eq}=1+\beta_n\omega_n\Delta t+\frac{\omega_n^2\Delta t^2}{6} \tag{21d}$$

with the acceleration $\ddot{Z}_n(t+\Delta t)$ determined, the corresponding velocity and displacement are determined from

$$\dot{Z}_n(t+\Delta t) = \dot{q}_n + \frac{\Delta t}{2}\ddot{Z}_n(t+\Delta t) \tag{22}$$

$$Z_n(t+\Delta t) = q_n + \frac{\Delta t^2}{6}\ddot{Z}_n(t+\Delta t) \tag{23}$$

For specified initial values of $Z_n(0)$ and $\dot{Z}_n(0)$ the initial acceleration, $\ddot{Z}_n(0)$ may be evaluated directly from Eq. (19), i.e.

$$\ddot{Z}_n(0) = F_n(0) - 2\beta_n\omega_n\dot{Z}_n(0) - \omega_n^2 Z_n(0) \tag{24}$$

with $Z_n(0)$, $\dot{Z}_n(0)$ and $\ddot{Z}_n(0)$ known, the values of the acceleration, velocity, and displacement at any time may be determined by repeated application of Eqs. (20), (22) and (23).

Once the values of Z_n have been determined, the deformations $\{U\}$ are determined from Eq. (18), and the deformations of the individual columns are determined from the following equations:

$$U_{ix} = U_x + \eta_i b U_\theta + U_{cx} \tag{25a}$$

$$U_{jy} = U_y - \xi_j a U_\theta + U_{cy} \tag{25b}$$

where

$$U_{cx} = \sum_{i=1}^{n}\left[k_{ix}^{*}\left(x_g\right)_i\right] - \left(x_g\right)_i + \eta_i T_\theta \tag{26a}$$

$$U_{cy} = \sum_{j=1}^{m}\left[k_{jy}^{*}\left(y_g\right)_j\right] - \left(y_g\right)_j - \xi_j T_\theta \tag{26b}$$

T_θ is the corresponding displacement term of $\ddot{T}_\theta$.

Note that the terms U_{cx} and U_{cy} in Eqs. (25) are related only to the input functions, the individual stiffnesses of the system, and the geometric arrangement of the columns. They account for the difference in the displacement values of the ground motion at the locations of the columns due to the finite speed of propagation of the excitation. This fact is normally ignored in the conventional analysis. Hereafter, both quantities U_{cx} and U_{cy} are designated as U_c the displacement correction term.

2.2.3 Equations for symmetric systems

The equations are applicable to both symmetric and eccentric systems. A symmetric system may experience torsional response even under the influence of a translational motion of the ground. This fact, which appears to have been first investigated by Newmark, can be seen clearly from Eqs. (14). In Newmark's approach, the rotational component of the response is first evaluated, and then it is combined with the translational component determined in the usual way. The method of combining the two components was not specified, (Newmark, 1969).

The method used herein consists of using Eqs. (14) with the e_x and e_y set equal to zero. In this approach, the torsional and translational effects are obtained simultaneously by solving the following set of equations:

$$\ddot{U}_x + 2\beta_x\omega_x\dot{U}_x + \omega_x^2 U_x = -\sum_{i=1}^{n} k_{ix}^{*}(\ddot{x}_o)_i \tag{27a}$$

$$\ddot{U}_y + 2\beta_y\omega_y\dot{U}_y + \omega_y^2 U_y = -\sum_{j=1}^{m} k_{jy}^{*}(\ddot{y}_o)_j \tag{27b}$$

$$\ddot{U}_\theta + 2\beta_\theta\omega_\theta\dot{U}_\theta + \omega_\theta^2 U_\theta = -\frac{1}{b}\ddot{T}_\theta \tag{27c}$$

Note that these equations are independent of one another. Accordingly, each unknown coordinate may be evaluated independently by the method described in the preceding section and the column deformations computed by use of Eqs. (25).

If the speed of propagation of the ground motion were infinite, as is normally assumed, the right-hand members of the three equations would be, respectively, $-\ddot{x}_o$, $-\ddot{y}_o$ and *0*, and no torsional response could develop. It is also interesting to note that, if only a single component of ground shaking is considered, say $\ddot{x}_o$ and if Eq. (27c) is multiplied by $b/2$, a direct comparison with the Newmark approach becomes possible.

Now, assume that the system has the same number of columns in the x and y directions and that the total stiffness of the system in either direction is equally distributed among the columns. The Newmark approach may be determined by neglecting the damping term in the Eq. (27c). For a very small transit time, t_r, which is the time required for the ground motion to traverse the longer plan dimension of the foundation, b, the Eq. (27c) may be written corresponding to the Newmark equation as follows,

$$-\frac{\ddot{x}_o t_r}{2}\left[1 \Big/ 1 + \frac{K_y}{K_x}\left(\frac{a}{b}\right)^2\right] \tag{28}$$

2.2.4 Spectral response

In the time history method, there have been two major disadvantages in the use of this approach. First, the method produces a large amount of output information that can require a significant amount of computational efforts to conduct all possible design checks, as a function of time. Second, the analysis must be repeated for several different earthquake motions in order to assure that all frequencies are excited, since a response spectrum for one earthquake in a specified direction is not a smooth function.

There are computational advantages in using the spectral response method of seismic analysis for prediction of displacements and member forces in structural systems. The method involves the calculation of only the maximum values of the displacements and member forces in each mode using smooth design spectrum that are the average of several ground motions.

If the ground motions in both x and y directions are characterized by the same response spectrum, then the maximum value of the modal response function (19) may be rewritten in the following form

$$|Z_n|_{\max} = \frac{1}{\omega_n^2}\left(\Gamma_{nx}a_x + \Gamma_{ny}a_y\right)S_a(\omega_n,\beta_n) \tag{29}$$

In which a_x and a_y are the amplitudes of the components of the ground motion along the x- and y- axes, respectively, and $S_a(\omega_n,\beta_n)$ is the spectral acceleration. Two Mode Participation Factors Γ_{nx} and Γ_{ny} may be obtained in the following form

$$\Gamma_{nx} = \varphi_{xn} \tag{30a}$$

$$\Gamma_{ny} = \varphi_{yn} \tag{30b}$$

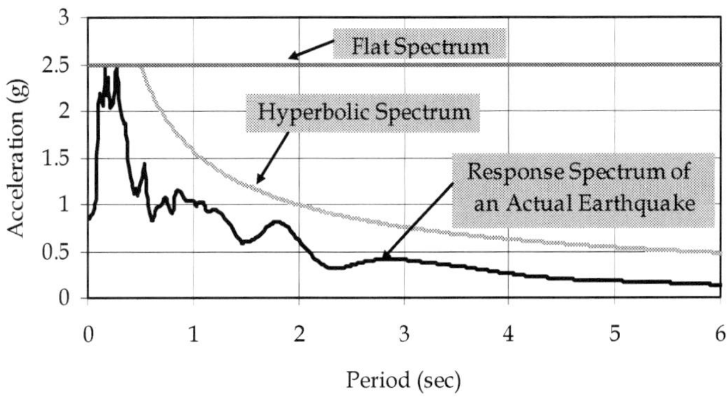

Fig. 12. Two idealized response spectrum

Finally, the most conservative method that is used to estimate a peak value of displacement or force within a structure is to use the sum of the absolute of the modal response values. This approach assumes that the maximum modal values, for all modes, occur at the same point in time. The relatively new method of modal combination is the Complete Quadratic Combination, CQC, method that was first published in 1981, (Wilson et al. 1981).

Now, the dynamic response of one-storey building model to the horizontal components of ground motion along x and y axes are investigated. For the objectives of this study it is considered the most appropriate to characterize ground motion by its response spectrum. The numerical results presented are for two idealized response spectrum (see Fig. 12), flat (or period independent) acceleration spectrum and hyperbolic acceleration spectrum (or flat velocity spectrum). The chosen spectrums have the advantage that the normalized response of the system does not depend on the periods of vibration, but only on their ratios.

For these two idealized spectrums, the normalized response does not depend on ω_n and ω_x separately but on the ratios ω_n/ω_x . The frequency ratios ω_n/ω_x and the mode shapes (φ_{xn} , φ_{xn} and $\varphi_{\theta n}$) depend on four dimensionless parameters (ω_y/ω_x , ω_θ/ω_x , ε_x and ε_y). The variation of the normalized modal responses, for a one-way coupled system ($e_x = 0$) subjected to ground motion in the x direction, for the case of flat spectrum, are shown in Fig. 13.

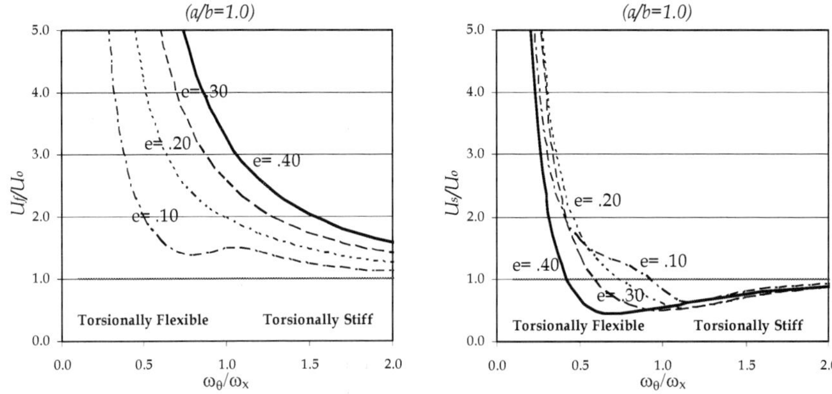

Fig. 13. Normalized displacements of the flexible and stiff edges; flat spectrum

Those for the case of hyperbolic spectrum are presented in Fig. 14. The maximum displacements of both flexible and stiff edges are calculated by modal superposition method, using complete CQC method. These values are then normalized by U_o the maximum displacement in the x direction produced by the same earthquake in an associated torsionally balanced building with stiffness and mass values similar to those of the asymmetric building but coincident centers of mass and rigidity.

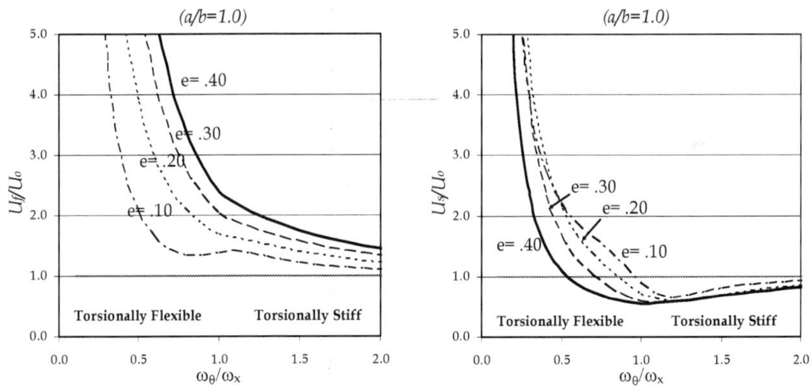

Fig. 14. Normalized displacements of the stiff and flexible edges; hyperbolic spectrum

The normalized flexible edge displacement U_f/U_o is plotted as a function of ω_θ/ω_x for different values of eccentricity $e = e_y$ and a plan aspect ratio of $a/b = 1$ in Fig. 11. In all cases flexible edge displacement in the structure is greater than the displacement of the associated symmetric structure. Of particular interest is the fact that there is a sharp increase in flexible edge displacement when ω_θ/ω_x falls below about 1.0.

It is also of interest to note that resonance between uncoupled translational and torsional frequencies, i.e., when $\omega_\theta/\omega_x = 1.0$, does not cause any significant increase in response.

Frequency resonance is not, therefore, a critical issue. Plots of normalized stiff edge displacement are shown in Fig. 13, again for different values of eccentricity $e = e_y$ and a plan aspect ratio of $a/b = 1$. Stiff edge displacement is less than 1 for $\omega_\theta/\omega_x \geq 1.0$. For $\omega_\theta/\omega_x < 1.0$, that is for torsionally flexible behaviour, stiff edge displacement starts to increase and can be, substantially, higher than 1. The results presented in Figs. 13 and 14 clearly suggest that buildings with low torsional stiffness may experience large displacements, causing distress in both structural and nonstructural components.

3. Torsional provisions in seismic codes as applied one-storey buildings

Most seismic building codes Formulate the design torsional moment at each storey as a product of the storey shear and a quantity termed design eccentricity. These provisions usually specify values of design eccentricities that are related to the static eccentricity between the center of rigidity and the center of mass. The earthquake-induced shears are applied through points located at the design eccentricities. A static analysis of the structure for such shears provides the design forces in the various elements of the structure. In some codes the design eccentricities include a multiplier on the static eccentricity to account for possible dynamic amplification of the torsion. The design eccentricities also include an allowance for accidental torsion. Such torsion is supposed to be induced by the rotational component of the ground motion and by possible deviation of the centers of rigidity and mass from their calculated positions. The design eccentricity formulae, given in building codes, can be written in two following parts:

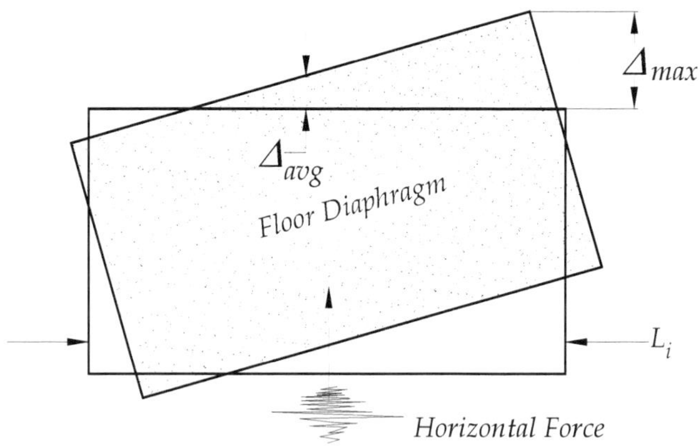

Fig. 15. Maximum and average diaphragm displacements of the structure

- The first part is expressed as some magnification factor times the structural eccentricity. This part deals with the complex nature of torsion and the effect of the simultaneous action of the two horizontal ground disturbances.
- The second term is called accidental eccentricity to account for the possible additional torsion arising from variations in the estimates of the relative rigidities, uncertain estimates of dead and live loads at the floor levels, addition of wall panels and partitions; after completion of the building, variation of the stiffness with time and, Inelastic or plastic action. The effects of possible torsional motion of the ground are also

considered to be included in this term. This terms in general a function of the plan dimension of the building in the direction of the computed eccentricity.

In Iranian code, in case of structures with rigid floors in their own plan, an additional accidental eccentricity is introduced through the effects generated by the uncertainties associate with the distribution of the mass level and/or the spatial variation of the ground seismic movement, (Iranian code 2800, 2005). This is considered for each design direction and for each level and also is related to the center of mass. The accidental eccentricity is computed with the relationship

$$e_i = \pm 0.05 L_i \tag{31}$$

where e_i is the accidental eccentricity of mass for storey *i* from its nominal location, applied in the same direction at all levels; L_i - the floor dimension perpendicular to the direction of the seismic action. If the lateral stiffness and mass are not distributed in plan and elevation, the accidental torsional effects may be accounted by multiplying an amplification factor A_j as follow

$$1.0 \le A_j = \left[\frac{\Delta_{\max}}{1.2\Delta_{avg}}\right]^2 \le 3.0 \tag{32}$$

where $\Delta_{\max}$ and Δ_{avg} are maximum and average diaphragm displacements of the structure, respectively, (see Fig. 15).

4. Conclusion

A study of free vibration characteristics of an eccentric one-storey structural model is presented. It is shown in the previous sections that the significance of the coupling effect of an eccentric system depends on the magnitude of the eccentricity between the centers of mass and of rigidity and the relative values of the uncoupled torsional and translational frequencies of the same system without taking the eccentricity into account. The coupling effect for a given eccentricity is the greatest when the uncoupled torsional frequency, ω_θ, and translational frequency, ω_x of the system are equal. As the value of ω_θ/ω_x increases, the coupling effect decreases. For small eccentricities, the motions may reasonably be considered uncoupled if the ratio of ω_θ/ω_x exceeds 2.5.

In addition, it is shown that the locus of the associated center of rotation can be formulated corresponding for a given eccentricity. Note that, for all values of eccentricity, as the value of the uncoupled natural frequencies ratio increases the center of rotation shifts away from the center of rigidity for the first mode and approaches the center of mass for the higher mode. It is also shown that, the torsional behaviour of the model assembled, using our approach, can be classified based on the nature of the instantaneous center of rotation.

It is well known that asymmetric or torsionally unbalanced buildings are vulnerable to damage during an earthquake. Resisting elements in such buildings could experience large displacements and distress. With eccentricity defined for one-storey buildings, the torsional provisions or building codes can then be applied for a seismic design or such structures.

5. Acknowledgment

The author gratefully acknowledges the financial support provided by the Office of Vice Chancellor for Research of Islamic Azad University, Kerman Branch.

6. References

9-11 Research Book, (2006). Other Building Collapses, Available from http://911research.wtc7.net/wtc/analysis/compare/collapses.html

Anastassiadis, K., Athanatopoulos, A. & Makarios, T. (1998). Equivalent static eccentricities in the simplified methods of seismic analysis of buildings, *Earthquake Spectra,* vol. 14, No. 1, pp.1–34.

Chandler, M. & Hutchinson, G.L. (1986). Torsional Coupling Effects in the Earthquake Response of Asymmetric Buildings, *Engineering Structures*, vol. 8, pp. 222-236.

Cruz, E.F. & Chopra, A.K. (1986). Simplified Procedures for Earthquake Analysis of Buildings, *Journal of Structural Engineering*, Vol. 112, pp. 461-480.

De la Llera, J.C. & Chopra, A.K. (1994). Using accidental eccentricity in code-specified static and dynamic analysis of buildings, *Earthquake Engineering and Structural Dynamics*, vol. 23, No. 7, pp. 947–967.

Dempsey, K.M. & Irvine, H.M. (1979). Envelopes of maximum seismic response for a partially symmetric single storey building model, *Earthquake Engineering and Structural Dynamics*, vol. 7, No. 2, pp. 161–180.

Earthquake Engineering ANNEXES, (2007), European Association for Earthquake Engineering.

Fajfar P., Marusic D. & Perus I. (2005). Torsional effects in the pushover-based seismic analysis of buildings. *Journal of Earthquake Engineering*, vol. 9, No. 6, pp. 831–854.

Hejal, R. & Chopra, A.K. (1989). Earthquake Analysis of a Class of Torsionally-Coupled Buildings, Earthquake Engineering and Structural Dynamics, Vol. 18, pp. 305-323.

Iranian Code of Practice for Seismic Resistant Design of Buildings. (2005). Standard No. 2800-05, 3rd Edition.

Koh, T., Takase, H. & Tsugawa, T. (1969). Torsional Problems in Seismic Design of High-Rise Buildings, *Proceedings of the Fourth World Conference on Earthquake Engineering*, Santiago, Chile, vol. 4, pp. 71-87.

Kuo, Pao-Tsin. (1974). Torsional Effects in Structures Subjected to Dynamic Excitations of the Ground, Ph.D. Thesis, Rice University.

Moghadam, A.S. & Tso, WK. (2000). Extension of Eurocode 8 torsional provisions to multi-storey buildings, *Journal of Earthquake Engineering*, vol. 4, No. 1, pp. 25–41.

Newmark, N. M., (1969). Torsion in Symmetrical Buildings, *Proceedings of the Fourth World Conference on Earthquake Engineering*, Vol. 2, Santiago, Chile, pp. A3-19 to A3-32.

Tabatabaei, R. & Saffari, H. (2010). Demonstration of Torsional Behaviour Using Vibration-based Single-storey Model with Double Eccentricities, *Journal of Civil Engineering*, vol. 14, No. 4., pp. 557-563.

Tanabashi, R. (1960). Non-Linear Transient Vibration of Structures, *Proceedings of the Second World Conference on Earthquake Engineering*, Tokyo, Japan, vol. 2, pp. 1223.

Tso, W.K. & Dempsey, K.M. (1980). Seismic torsional provisions for dynamic eccentricity, *Earthquake Engineering and Structural Dynamics*, vol. 8, No. 3, pp. 275–289.

Wilson, E. L., Der Kiureghian, A. & Bayo, E. R. (1981). A Replacement for the SRSS Method in Seismic Analysis, *Earthquake Engineering and Structural Dynamics*, Vol. 9, pp. 187-192.

10

Beam Structural Modelling in Hydroelastic Analysis of Ultra Large Container Ships

Ivo Senjanović, Nikola Vladimir, Neven Hadžić and Marko Tomić
University of Zagreb, Faculty of Mechanical Engineering and Naval Architecture
Croatia

1. Introduction

Ultra large container ships are very sensitive to the wave load of quartering seas due to considerably reduced torsional stiffness caused by large deck openings. As a result, their natural frequencies can fall into the range of encounter frequencies in an ordinary sea spectrum. Therefore, the wave induced hydroelastic response of large container ships becomes an important issue in structural design. Mathematical hydroelastic model incorporates structural, hydrostatic and hydrodynamic parts (Senjanović et al. 2007, 2008a, 2009b, 2010b). Beam structural model is preferable in the early design stage and for determining global response, while for more detailed analyses 3D FEM model has to be used. The hydroelastic analysis is performed by the modal superposition method, which requires dry natural vibrations of the structure to be determined. For each mode dynamic coefficients (added mass and damping) and wave load are calculated based on velocity potential. The governing equation of ship motion in rough sea specified for the impulsive (slamming) load as a transient problem is solved in time domain. The motion equation is also given for the case of harmonic wave excitation (springing), which is solved in the frequency domain.

In the chapter, methodology of the ship hydroelastic analysis is described, and position and role of the beam structural model is explained. Beam finite element for coupled horizontal and torsional vibrations, that includes warping of ship cross-section, is constructed. Shear influence on both bending and torsion is taken into account. The strip element method is used for determination of normal and shear stress flows, and stiffness moduli, i.e. shear area, torsional modulus, shear inertia modulus (as a novelty), and warping modulus.

In the modelling of large container ships it is important to appropriately account for the contribution of transverse bulkheads to hull stiffness and the behavior of relatively short engine room structure. In the former case, the equivalent torsional modulus is determined by increasing ordinary (St. Venant) value, depending on the ratio of the strain energy of a bulkhead and corresponding hull portion. Equivalent torsional modulus of the engine room structure is also determined utilizing the energy approach. It is assumed that a short closed structure behaves as an open one with the contribution of decks.

Application of the beam structural model for ship hydroelastic analysis is illustrated in case of a very large container ship. Correlation of dry natural vibrations analysis results for the beam model with those for 3D FEM model shows very good agreement. Hydroelastic analysis emphasizes peak values of transfer functions of displacements and sectional forces in resonances, i.e. in the case when the encounter frequency is equal to one of the natural frequencies.

2. Methodology of ship hydroelastic analysis

A structural model, ship and cargo mass distributions and geometrical model of ship surface have to be defined to perform ship hydroelastic analysis. At the beginning, dry natural vibrations have to be calculated, and after that modal hydrostatic stiffness, modal added mass, damping and modal wave load are determined. Finally, wet natural vibrations as well as the transfer functions (RAO) for determining ship structural response to wave excitation are obtained (Senjanović et al. 2008a, 2009b), Fig. 1.

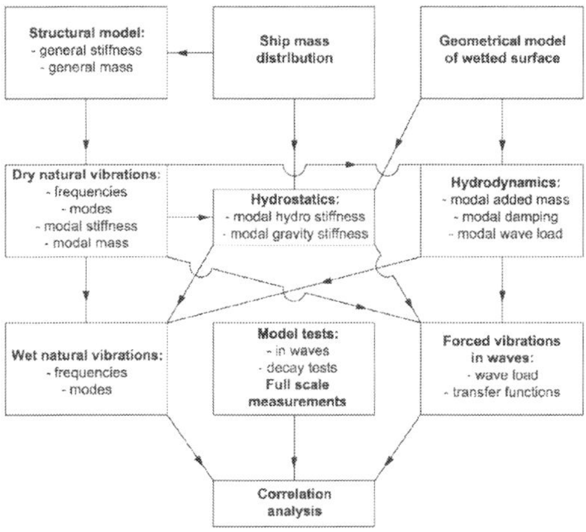

Fig. 1. Methodology of ship hydroelastic analysis

3. General remarks on structural model

A ship hull, as an elastic non-prismatic thin-walled girder, performs longitudinal, vertical, horizontal and torsional vibrations. Since the cross-sectional centre of gravity and centroid, as well as the shear centre positions are not identical, coupled longitudinal and vertical, and horizontal and torsional vibrations occur, respectively. The shear centre in ships with large hatch openings is located below the keel and therefore the coupling of horizontal and torsional vibrations is extremely high. The above problem is rather complex due to geometrical discontinuity of the hull cross-section, Fig. 2.

The accuracy of the solution depends on the reliability of stiffness parameters determination, i.e. of bending, shear, torsional and warping moduli. The finite element method is a powerful tool to solve the above problem in a successful way. One of the first solutions for coupled horizontal and torsional hull vibrations, dealing with the finite element technique, is given in (Kawai, 1973, Senjanović & Grubišić, 1991). Generalised and improved solutions are presented in (Pedersen, 1985, Wu & Ho, 1987). In all these references, the determination of hull stiffness is based on the classical thin-walled girder theory, which doesn't give a satisfactory value for the warping modulus of the open cross-section (Haslum & Tonnessen, 1972, Vlasov, 1961). Apart from that, the fixed values of stiffness moduli are determined, so that the application of the beam theory for hull vibration analysis is limited to a few lowest natural modes only. Otherwise, if the mode dependent stiffness parameters are used the application of the beam theory can be extended up to the tenth natural mode (Senjanović & Fan, 1989, 1992, 1997).

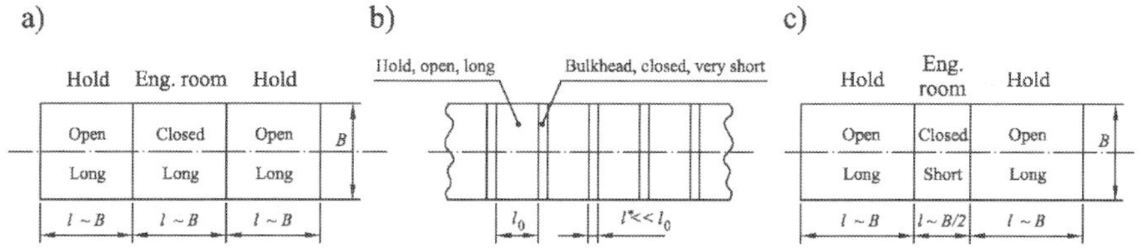

Fig. 2. Discontinuities of ship hull

4. Consistent differential equations of beam vibrations

Referring to the flexural beam theory (Timoshenko & Young, 1955, Senjanović, 1990), the total beam deflection, w, consists of the bending deflection, w_b, and the shear deflection, w_s, while the angle of cross-section rotation depends only on the former, Fig. 3

$$w = w_b + w_s, \quad \varphi = \frac{\partial w_b}{\partial x} \tag{1}$$

The cross-sectional forces are the bending moment and the shear force

$$M = -EI_b \frac{\partial \varphi}{\partial x}, \tag{2}$$

$$Q = GA_s \frac{\partial w_s}{\partial x}, \tag{3}$$

where E and G are the Young's and shear modulus, respectively, while I_b and A_s are the moment of inertia of cross-section and shear area, respectively.
The inertia load consists of the distributed transverse load, q_i, and the bending moment, μ_i, and in the case of coupled horizontal and torsional vibration is specified as

$$q_i = -m\left(\frac{\partial^2 w}{\partial t^2} + c\frac{\partial^2 \psi}{\partial t^2}\right), \tag{4}$$

$$\mu_i = -J_b \frac{\partial^2 \varphi}{\partial t^2}, \tag{5}$$

where m is the distributed mass, J_b is the mass moment of inertia about z-axis, and c is the distance between the centre of gravity and the shear centre, $c = z_G - z_S$, Fig. 4.

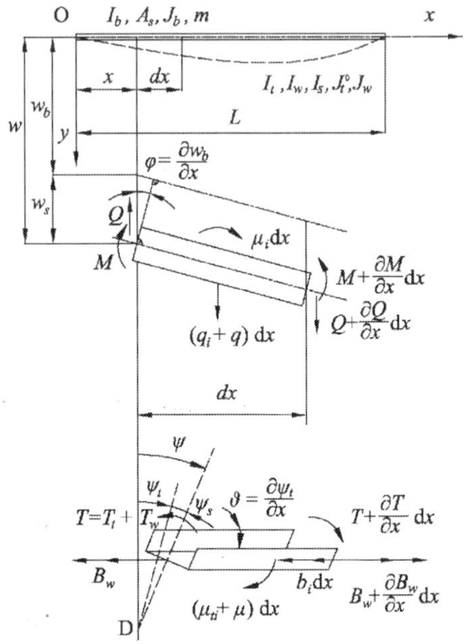

Fig. 3. Beam bending and torsion

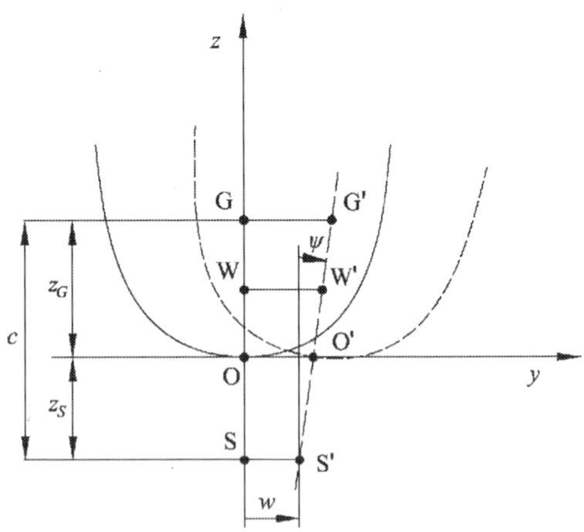

Fig. 4. Cross-section of a thin-walled girder

In a similar way the total twist angle, ψ, consists of the pure twist angle, ψ_t, and the shear contribution, ψ_s, while the second torsional displacement, which causes warping of cross-section, is variation of the pure twist angle, i.e. Fig. 3 (Pavazza, 2005)

$$\psi = \psi_t + \psi_s, \quad \vartheta = \frac{\partial \psi_t}{\partial x}. \tag{6}$$

The cross-sectional forces include the pure torsional torque, T_t, warping bimoment, B_w, and additional torque due to restrained warping, T_w

$$T_t = GI_t \vartheta, \tag{7}$$

$$B_w = -EI_w \frac{\partial \vartheta}{\partial x}, \tag{8}$$

$$T_w = GI_s \frac{\partial \psi_s}{\partial x}, \tag{9}$$

where I_t, I_w and I_s are the torsional modulus, warping modulus and shear inertia modulus, respectively.

The inertia load consists of the distributed torque, μ_{ti}, and the bimoment, b_i, presented in the following form:

$$\mu_{ti} = -J_t \frac{\partial^2 \psi}{\partial t^2} - mc \frac{\partial^2 w}{\partial t^2}, \tag{10}$$

$$b_i = -J_w \frac{\partial^2 \vartheta}{\partial t^2}, \tag{11}$$

where J_t is the mass polar moment of inertia about the shear centre, and J_w is the mass bimoment of inertia with respect to the warping centre, Fig. 4.

Considering the equilibrium of a beam differential element, one can write for flexural vibrations

$$\frac{\partial M}{\partial x} = Q + \mu_i, \tag{12}$$

$$\frac{\partial Q}{\partial x} = -q_i - q, \tag{13}$$

and for torsional vibrations (Pavazza, 1991)

$$\frac{\partial B_w}{\partial x} = T_w + b_i, \tag{14}$$

$$\frac{\partial T_t}{\partial x} + \frac{\partial T_w}{\partial x} = -\mu_{ti} - \mu. \tag{15}$$

The above equations can be reduced to two coupled partial differential equations as follows. Substituting Eqs. (2) and (3) into (12) yields

$$\frac{\partial w_s}{\partial x} = -\frac{EI_b}{GA_s}\frac{\partial^2 \varphi}{\partial x^2} + \frac{J_b}{GA_s}\frac{\partial^2 \varphi}{\partial t^2}. \tag{16}$$

By inserting Eqs. (3) and (4) into (13) leads to

$$EI_b\frac{\partial^4 \varphi}{\partial x^4} + m\frac{\partial^2 \varphi}{\partial t^2} - \left(J_b + m\frac{EI_b}{GA_s}\right)\frac{\partial^4 \varphi}{\partial x^2 \partial t^2} + \frac{mJ_b}{GA_s}\frac{\partial^4 \varphi}{\partial t^4} + mc\frac{\partial^3 \psi}{\partial x \partial t^2} = \frac{\partial q}{\partial x}. \tag{17}$$

In a similar way, substituting Eqs. (8) and (9) into (14) yields

$$\frac{\partial \psi_s}{\partial x} = -\frac{EI_w}{GI_s}\frac{\partial^2 \vartheta}{\partial x^2} + \frac{J_w}{GI_s}\frac{\partial^2 \vartheta}{\partial t^2}. \tag{18}$$

By inserting Eqs. (7), (9) and (10) into (15) one finds

$$EI_w\frac{\partial^4 \vartheta}{\partial x^4} - GI_t\frac{\partial^2 \vartheta}{\partial x^2} + J_t\frac{\partial^2 \vartheta}{\partial t^2} - \left(J_w + J_t\frac{EI_w}{GI_s}\right)\frac{\partial^4 \vartheta}{\partial x^2 \partial t^2} + \frac{J_w}{GI_s}\frac{\partial^4 \vartheta}{\partial t^4} + mc\frac{\partial^3 w}{\partial x \partial t^2} = \frac{\partial \mu}{\partial x}. \tag{19}$$

Furthermore, ψ in (17) can be split into $\psi_t + \psi_s$ and the later term can be expressed with (18). Similar substitution can be done for $w = w_b + w_s$ in (19), where w_s is given with (16). Thus, taking into account that $\varphi = \partial w_b / \partial x$ and $\vartheta = \partial \psi_t / \partial x$, Eqs. (17) and (19) after integration per x read

$$\begin{aligned} &EI_b\frac{\partial^4 w_b}{\partial x^4} + m\frac{\partial^2 w_b}{\partial t^2} - \left(J_b + m\frac{EI_b}{GA_s}\right)\frac{\partial^4 w_b}{\partial x^2 \partial t^2} + \frac{mJ_b}{GA_s}\frac{\partial^4 w_b}{\partial t^4} \\ &+ mc\left(\frac{\partial^2 \psi_t}{\partial t^2} - \frac{EI_w}{GI_s}\frac{\partial^4 \psi_t}{\partial x^2 \partial t^2} + \frac{J_w}{GI_s}\frac{\partial^4 \psi_t}{\partial t^4}\right) = q \end{aligned} \tag{20}$$

$$\begin{aligned} &EI_w\frac{\partial^4 \psi_t}{\partial x^4} - GI_t\frac{\partial^2 \psi_t}{\partial x^2} + J_t\frac{\partial^2 \psi_t}{\partial t^2} - \left(J_w + J_t\frac{EI_w}{GI_s}\right)\frac{\partial^4 \psi_t}{\partial x^2 \partial t^2} \\ &+ \frac{J_w}{GI_s}\frac{\partial^4 \psi_t}{\partial t^4} + mc\left(\frac{\partial^2 w_b}{\partial t^2} - \frac{EI_b}{GA_s}\frac{\partial^4 w_b}{\partial x^2 \partial t^2} + \frac{J_b}{GA_s}\frac{\partial^4 w_b}{\partial t^4}\right) = \mu. \end{aligned} \tag{21}$$

After solving Eqs. (20) and (21) the total deflection and twist angle are obtained by employing (16) and (18), i.e.

$$w = w_b + w_s = w_b - \frac{EI_b}{GA_s}\frac{\partial^2 w_b}{\partial x^2} + \frac{J_b}{GA_s}\frac{\partial^2 w_b}{\partial t^2} + f(t) \tag{22}$$

$$\psi = \psi_t + \psi_s = \psi_t - \frac{EI_w}{GI_s}\frac{\partial^2 \psi_t}{\partial x^2} + \frac{J_w}{GI_s}\frac{\partial^2 \psi_t}{\partial t^2} + g(t) \tag{23}$$

where $f(t)$ and $g(t)$ are integration functions, which depend on initial conditions.
The main purpose of developing differential equations of vibrations (20) and (21) is to get insight into their constitution, position and role of the stiffness and mass parameters, and

coupling, which is realized through the inertia terms. If the pure torque T_t is excluded from the above theoretical consideration, it is obvious that the complete analogy between bending and torsion exists, (Pavazza, 1991).

Application of Eqs. (20) and (21) is limited to prismatic girders. For more complex problems, like ship hull, the finite element method is on disposal.

The shape functions of beam finite element for vibration analysis have to satisfy the following consistency relations for harmonic vibrations obtained from Eqs. (22) and(23), (Senjanović, 1990)

$$w = w_b + w_s = \left(1 - \omega^2 \frac{J_b}{GA_s}\right) w_b - \frac{EI_b}{GA_s} \frac{\mathrm{d}^2 w_b}{\mathrm{d}x^2} \tag{24}$$

$$\psi = \psi_t + \psi_s = \left(1 - \omega^2 \frac{J_w}{GI_s}\right) \psi_t - \frac{EI_w}{GI_s} \frac{\mathrm{d}^2 \psi_t}{\mathrm{d}x^2}. \tag{25}$$

5. Beam finite element

The properties of a finite element for the coupled horizontal and torsional vibration analysis can be derived from the total element energy. It consists of the strain energy, the kinetic energy, the work of the distributed external lateral load, q, and the torque, μ, and the work of the boundary forces. Thus, according to (Senjanović, 1990, Senjanović & Grubišić, 1991),

$$\begin{aligned} E_{tot} &= \frac{1}{2}\int_0^l \left[EI_b \left(\frac{\partial^2 w_b}{\partial x^2}\right)^2 + GA_s \left(\frac{\partial w_s}{\partial x}\right)^2 + EI_w \left(\frac{\partial^2 \psi_t}{\partial x^2}\right)^2 + GI_s \left(\frac{\partial \psi_s}{\partial x}\right)^2 + GI_t \left(\frac{\partial \psi_t}{\partial x}\right)^2 \right] \mathrm{d}x \\ &+ \frac{1}{2}\int_0^l \left[m \left(\frac{\partial w}{\partial t}\right)^2 + J_b \left(\frac{\partial^2 w_b}{\partial x \partial t}\right)^2 + 2mc \frac{\partial w}{\partial t}\frac{\partial \psi}{\partial t} + J_w \left(\frac{\partial^2 \psi_t}{\partial x \partial t}\right)^2 + J_t \left(\frac{\partial \psi}{\partial t}\right)^2 \right] \mathrm{d}x \\ &- \int_0^l (qw + \mu\psi)\mathrm{d}x + (Qw - M\varphi + T\psi + B_w \vartheta)_0^l , \end{aligned} \tag{26}$$

where l is the element length. Since the beam has four displacements, $w, \varphi, \psi, \vartheta$, a two-node finite element has eight degrees of freedom, i.e. four nodal shear-bending and torsion-warping displacements respectively, Fig. 5,

$$\{U\} = \begin{Bmatrix} w(0) \\ \varphi(0) \\ w(l) \\ \varphi(l) \end{Bmatrix}, \quad \{V\} = \begin{Bmatrix} \psi(0) \\ \vartheta(0) \\ \psi(l) \\ \vartheta(l) \end{Bmatrix} \tag{27}$$

Therefore, the basic beam displacements, w_b and ψ_t, can be presented as the third-order polynomials

$$\begin{aligned} &w_b = \langle a_k \rangle \{\xi^k\}, \ \psi_t = \langle d_k \rangle \{\xi^k\}, \ k = 0,1,2,3, \\ &\xi = \frac{x}{l}, \ \langle \ldots \rangle = \{ \ldots \}^T . \end{aligned} \tag{28}$$

Furthermore, satisfying alternately the unit value for one of the nodal displacement $\{U\}$ and zero values for the remaining displacements, and doing the same for $\{V\}$, it follows that:

$$w_b = \langle w_{bi} \rangle \{U\},\ w_s = \langle w_{si} \rangle \{U\},\ w = \langle w_i \rangle \{U\},$$
$$\psi_t = \langle \psi_{ti} \rangle \{V\},\ \psi_s = \langle \psi_{si} \rangle \{V\},\ \psi = \langle \psi_i \rangle \{V\},\ i = 1,2,3,4, \tag{29}$$

where w_{bi}, w_{si}, w_i and ψ_{ti}, ψ_{si}, ψ_i are the shape functions specified below by employing relations (24) and (25)

$$w_{bi} = \langle a_{ik} \rangle \{\xi^k\},\ w_{si} = \langle b_{ik} \rangle \{\xi^k\},\ w_i = \langle c_{ik} \rangle \{\xi^k\}$$
$$\psi_{ti} = \langle d_{ik} \rangle \{\xi^k\},\ \psi_{si} = \langle e_{ik} \rangle \{\xi^k\},\ \psi_i = \langle f_{ik} \rangle \{\xi^k\} \tag{30}$$

$$[a_{ik}] = \frac{1}{a(a+12\beta)} \begin{bmatrix} a+6\beta & 0 & -3a & 2a \\ -4\beta(a+3\beta)l & a(a+12\beta)l & -2a(a+3\beta)l & a^2 l \\ 6\beta & 0 & 3a & -2a \\ -2\beta(a-6\beta)l & 0 & -a(a-6\beta)l & a^2 l \end{bmatrix} \tag{31}$$

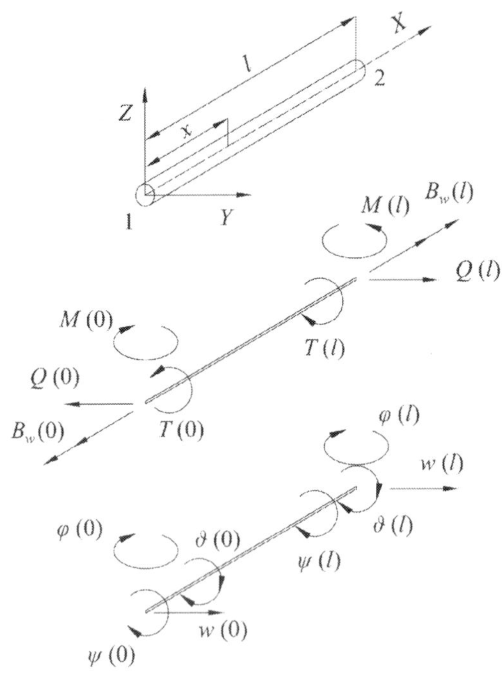

Fig. 5. Beam finite elemet

$$\begin{aligned} b_{i0} &= -(1-a)a_{i0} - 2\beta a_{i2} \\ b_{i1} &= -(1-a)a_{i1} - 6\beta a_{i3} \\ b_{i2} &= -(1-a)a_{i2} \\ b_{i3} &= -(1-a)a_{i3}, \end{aligned} \tag{32}$$

$$\left[c_{ik}\right]=\left[a_{ik}\right]+\left[b_{ik}\right], \quad i=1,2,3,4, \quad k=0,1,2,3 \tag{33}$$

$$\alpha = 1-\omega^2 \frac{J_b}{GA_s}, \ \beta = \frac{EI_b}{GA_s l^2} \tag{34}$$

Constitution of torsional matrices $\left[d_{ik}\right]$, $\left[e_{ik}\right]$ and $\left[f_{ik}\right]$ is the same as $\left[a_{ik}\right]$, $\left[b_{ik}\right]$ and $\left[c_{ik}\right]$, but parameters a and β have to be exchanged with

$$\eta = 1-\omega^2 \frac{J_w}{GI_s}, \ \gamma = \frac{EI_w}{GI_s l^2} \tag{35}$$

according to (25). By substituting Eqs. (29) (26) one obtains

$$E_{tot} = \frac{1}{2}\begin{Bmatrix} U \\ V \end{Bmatrix}^T \begin{bmatrix} k_{bs} & 0 \\ 0 & k_{ws}+k_t \end{bmatrix} \begin{Bmatrix} U \\ V \end{Bmatrix} + \frac{1}{2}\begin{Bmatrix} \dot{U} \\ \dot{V} \end{Bmatrix}^T \begin{bmatrix} m_{sb} & m_{st} \\ m_{ts} & m_{tw} \end{bmatrix} \begin{Bmatrix} \dot{U} \\ \dot{V} \end{Bmatrix} - \begin{Bmatrix} q \\ \mu \end{Bmatrix}^T \begin{Bmatrix} U \\ V \end{Bmatrix} - \begin{Bmatrix} P \\ R \end{Bmatrix}^T \begin{Bmatrix} U \\ V \end{Bmatrix}, \tag{36}$$

where, assuming constant values of the element properties,

$$\left[k\right]_{bs} = \left[EI_b \int_0^l \frac{\mathrm{d}^2 w_{bi}}{\mathrm{d}x^2}\frac{\mathrm{d}^2 w_{bj}}{\mathrm{d}x^2}\mathrm{d}x + GA_s \int_0^l \frac{\mathrm{d}w_{si}}{\mathrm{d}x}\frac{\mathrm{d}w_{sj}}{\mathrm{d}x}\mathrm{d}x \right]$$ - bending-shear stiffness matrix,

$$\left[k\right]_{ws} = \left[EI_w \int_0^l \frac{\mathrm{d}^2 \psi_{ti}}{\mathrm{d}x^2}\frac{\mathrm{d}^2 \psi_{tj}}{\mathrm{d}x^2}\mathrm{d}x + GI_s \int_0^l \frac{\mathrm{d}\psi_{si}}{\mathrm{d}x}\frac{\mathrm{d}\psi_{sj}}{\mathrm{d}x}\mathrm{d}x \right]$$ - warping-shear stiffness matrix,

$$\left[k\right]_{t} = \left[GI_t \int_0^l \frac{\mathrm{d}\psi_{ti}}{\mathrm{d}x}\frac{\mathrm{d}\psi_{tj}}{\mathrm{d}x}\mathrm{d}x \right]$$ - torsion stiffness matrix,

$$\left[m\right]_{sb} = \left[m\int_0^l w_i w_j \mathrm{d}x + J_b \int_0^l \frac{\mathrm{d}w_{bi}}{\mathrm{d}x}\frac{\mathrm{d}w_{bj}}{\mathrm{d}x}\mathrm{d}x \right]$$ - shear-bending mass matrix,

$$\left[m\right]_{tw} = \left[J_t\int_0^l \psi_i \psi_j \mathrm{d}x + J_w \int_0^l \frac{\mathrm{d}\psi_{ti}}{\mathrm{d}x}\frac{\mathrm{d}\psi_{tj}}{\mathrm{d}x}\mathrm{d}x \right]$$ - torsion-warping mass matrix,

$$\left[m\right]_{st} = \left[mc\int_0^l w_i \psi_j \mathrm{d}x \right], \quad \left[m\right]_{ts} = \left[m\right]_{st}^T$$ - shear-torsion mass matrix,

$$\begin{aligned} \{q\} &= \left\{ \int_0^l q w_j \,\mathrm{d}x \right\} \text{ - shear load vector,} \\ \{\mu\} &= \left\{ \int_0^l \mu \psi_j \,\mathrm{d}x \right\} \text{ - torsion load vector,} \\ & i,j = 1,2,3,4. \end{aligned} \tag{37}$$

The vectors $\{P\}$ and $\{R\}$ in Eq. (36) represent the shear-bending and torsion-warping nodal forces, respectively,

$$\{P\} = \begin{Bmatrix} -Q(0) \\ M(0) \\ Q(l) \\ -M(l) \end{Bmatrix}, \quad \{R\} = \begin{Bmatrix} -T(0) \\ B_w(0) \\ T(l) \\ -B_w(l) \end{Bmatrix} \tag{38}$$

The above matrices are specified in Appendix A, as well as the load vectors for linearly distributed loads along the finite element, i.e.

$$q = q_0 + q_1\xi, \quad \mu = \mu_0 + \mu_1\xi. \tag{39}$$

The total element energy has to be at its minimum. Satisfying the relevant conditions

$$\frac{\partial E_{tot}}{\partial\{U\}} = \{0\}, \quad \frac{\partial E_{tot}}{\partial\{V\}} = \{0\} \tag{40}$$

and employing Lagrange equations of motion, the finite element equation yields

$$\{f\} = [k]\{\delta\} + [m]\{\ddot{\delta}\} - \{f\}_{q\mu}, \tag{41}$$

where

$$\{f\} = \begin{Bmatrix} P \\ R \end{Bmatrix}, \quad \{f\}_{q\mu} = \begin{Bmatrix} q \\ \mu \end{Bmatrix}, \quad \{\delta\} = \begin{Bmatrix} U \\ V \end{Bmatrix}$$
$$[k] = \begin{bmatrix} k_{bs} & 0 \\ 0 & k_{ws} + k_t \end{bmatrix}, \quad [m] = \begin{bmatrix} m_{sb} & m_{st} \\ m_{ts} & m_{tw} \end{bmatrix}. \tag{42}$$

It is obvious that coupling between the bending and torsion occurs through the mass matrix only, i.e. by the coupling matrices $[m]_{st}$ and $[m]_{ts}$.

6. Contribution of transverse bulkheads to the hull stiffness

This problem for container ships is extensively analyzed in (Senjanović et al., 2008b), where torsional modulus of ship cross-section is increased proportionally to the ratio of bulkhead strain energy and strain energy of corresponding hull portion. The bulkhead is considered as an orthotropic plate with very strong stool (Szilard, 2004). Bulkhead strain energy is determined for the given warping of cross-section as a boundary condition. The warping causes bulkhead screwing and bending. Here, only the review of the final results is presented. Bulkhead deflection (axial displacement) is given by the following formula, Fig. 6:

$$u(y,z) = -y\left\{(z-d) + \left[1 - \left(\frac{y}{b}\right)^2\right]\frac{z^2}{H}\left(2 - \frac{z}{H}\right)\right\}\psi', \tag{43}$$

where H is the ship height, b is one half of bulkhead breadth, d is the distance of warping centre from double bottom neutral line, y and z are transverse and vertical coordinates, respectively, and ψ' is the variation of twist angle.

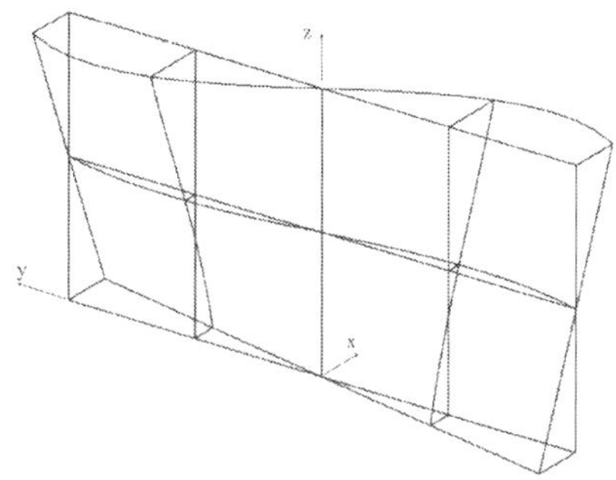

Fig. 6. Shape of bulkhead deformation

The bulkhead grillage strain energy includes vertical and horizontal bending with contraction and torsion (Senjanović et al., 2008b).

$$U_g = \frac{1}{1-\nu^2}\left[\frac{116H^3}{35b}i_y + \frac{32b^3}{105H}i_z + \frac{8Hb}{75}\nu\left(i_y + i_z\right) + \frac{143Hb}{75}\left(1-\nu\right)i_t\right]E\psi'^2 \tag{44}$$

where i_y, i_z and i_t are the average moments of inertia of cross-section and torsional modulus per unit breadth, respectively. The stool strain energy is comprised of the bending, shear and torsional contributions

$$U_s = \left[\frac{12h^2 I_{sb}}{b} + 72\left(1+\nu\right)\frac{h^2}{b^3}\frac{I_{sb}^2}{A_s} + \frac{9bI_{st}}{10\left(1+\nu\right)}\right]E\psi'^2 \tag{45}$$

where I_{sb}, A_s and I_{st} are the moment of inertia of cross-section, shear area and torsional modulus, respectively. Quantity h is the stool distance from the inner bottom, Fig. 7.

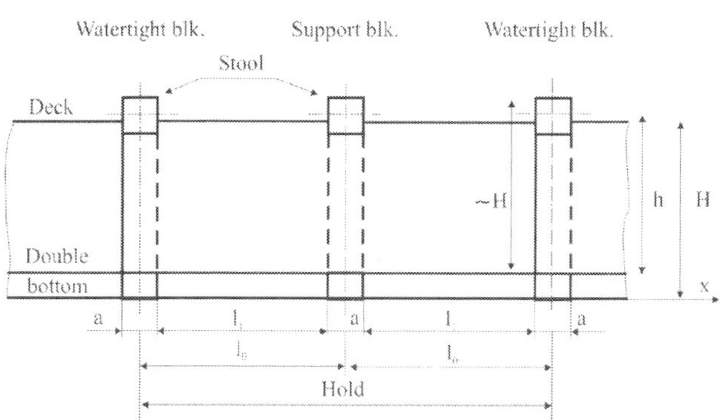

Fig. 7. Longitudinal section of container ship hold

The equivalent torsional modulus yields, Fig. 7

$$I_t^* = \left[1 + \frac{a}{l_1} + \frac{4(1+\nu)C}{I_t l_0}\right] I_t, \quad C = \frac{U_g + U_s}{E\psi'^2}, \tag{46}$$

where a is the web height of bulkhead girders (frame spacing), l_0 is the bulkhead spacing, $l_1 = l_0 - a$ is the net length, and C is the energy coefficient.

7. Contribution of engine room structure to the hull stiffness

Ultra Large Container Ships are characterized by relatively short engine room structure with length of about a half of ship breadth. Its complex deformation is illustrated in a case of a 7800 TEU container ship, Fig. 8. The deck shear deformation is predominant, while hold transverse bulkhead stool is exposed to bending. Due to shortness of the engine room, its transverse bulkheads are skewed but somewhat less pronounced than warping of the hold bulkheads. Warping of the transom is negligible, and that is an important fact when specifying boundary conditions in vibration analysis.

Fig. 8. Deformation of 7800 TEU container ship aft structure

7.1 Stiffness of engine room structure

A short engine room structure can be considered either as a closed segment with relevant stiffness or as an open segment with increased stiffness due to deck contribution (Pedersen, 1985). The latter simulation in fact gives results which agree better with 3D FEM results, than the former one (Pedersen, 1983). Deck contribution to hull stiffness can be determined by energy approach, as it is done in the case of transverse bulkheads (Senjanović et al., 2008b). Such a beam model is consistent at global level of energy balance, and that is sufficient for application in ship hydroelastic analysis, where proper natural frequencies and mode shapes of dry hull are required.

In the case of short engine room, torsion induces distortion of cross-section while hull bending is negligible. Solution of that complex problem is described here by employing the energy balance approach and concept of the effective stiffness due to reason of simplicity. A

closed hull segment is considered as open one with deck influence. For that purpose let us determine deck strain energy. All quantities related to closed and open cross-section are designated by $(.)^*$ and $(.)^\circ$, respectively
As it can be seen in Fig. 8, the upper deck is exposed to large deformation, while the double bottom in-plane deformation is quite small. The relative axial displacement of the internal upper deck boundaries, with respect to double bottom, is result of their warping

$$U = |U_D| + |U_B| = (|w_D| + |w_B|)\psi_t' \tag{47}$$

It causes deck in-plane (membrane) deformation. The problem can be solved in an approximate analytical way by considering deck as a beam. Its horizontal anti-symmetric deflection consists of pure bending and shear contribution, Fig. 9. The former is assumed in the form

$$u_b = \frac{y}{2b}\left[3 - \left(\frac{y}{b}\right)^2\right]U_b, \tag{48}$$

which satisfies relevant boundary conditions: $u_b(0) = 0$ and $u_b''(0) = 0$, where U_b is the boundary bending deflection. Shear deflection depends on bending deflection

$$u_s = -\frac{EI}{GA}\frac{\mathrm{d}^2 u_b}{\mathrm{d}y^2} = 2(1+\nu)\left(\frac{a}{b}\right)^2 \frac{y}{b}U_b, \tag{49}$$

where the internal deck cross-section area, $A = 2at$, its moment of inertia, $I = \frac{2}{3}a^3 t$, and the relation $E = 2(1+\nu)G$, are taken into account, Fig. 9. Total deflection is obtained by summing up its constitutive parts, Eqs. (48) and (49). Relation between total boundary deflection and the bending boundary deflection reads

$$U = \left[1 + 2(1+\nu)\left(\frac{a}{b}\right)^2\right]U_b \tag{50}$$

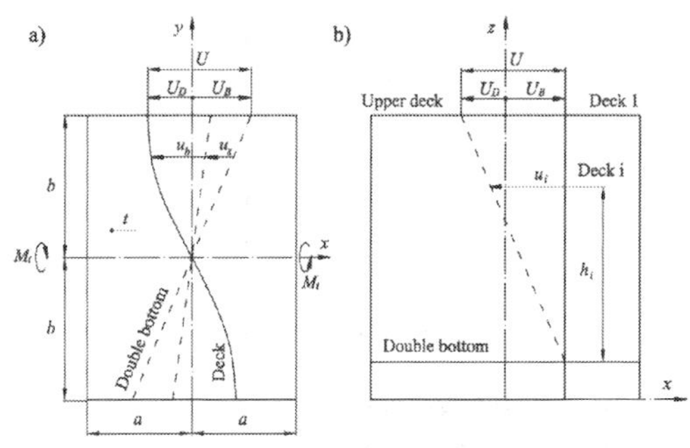

Fig. 9. Deck deformation and double bottom rotation, a)-bird view, b)-lateral view

The total internal deck strain energy consists of the bending and shear contributions

$$E_1 = \frac{1}{2} EI \int_{-b}^{b} \left(\frac{\mathrm{d}^2 u_b}{\mathrm{d}y^2} \right)^2 \mathrm{d}y + \frac{1}{2} GA \int_{-b}^{b} \left(\frac{\mathrm{d}u_s}{\mathrm{d}y} \right)^2 \mathrm{d}y \tag{51}$$

By substituting Eqs. (48) and (49) into (51), one finds

$$E_1 = 4(1+\nu) Gt \left(\frac{a}{b} \right)^3 \left[1 + 2(1+\nu) \left(\frac{a}{b} \right)^2 \right] U_b^2 \tag{52}$$

Finally, by taking into account Eqs. (47) and (50), yields

$$E_1 = \frac{4(1+\nu) Gt \left(\frac{a}{b} \right)^3}{1 + 2(1+\nu) \left(\frac{a}{b} \right)^2} \left(|w_D| + |w_B| \right)^2 \psi_t'^2 \tag{53}$$

On the other hand, total energy of the closed hull segment can be obtained by summing up energy of open segment and the deck strain energy, i.e.

$$E_{tot}^* = E_w^\circ + E_t^\circ + E_1 - E_\mu \tag{54}$$

where

$$E_w^\circ = \frac{1}{2} \int_{-a}^{a} B_w^\circ \psi_t'' \mathrm{d}x, \quad E_t^\circ = \frac{1}{2} \int_{-a}^{a} T_t^\circ \psi_t' \mathrm{d}x, \quad E_\mu = \int_{-a}^{a} \mu_x \psi \mathrm{d}x. \tag{55}$$

Within a short span $2a$, constant value of ψ_t' (as for deck) can be assumed, so that second term in Eq. (26) by inserting T_t° from Eqs. (7), leads to

$$E_t^\circ = GI_t^\circ a \psi_t'^2. \tag{56}$$

E_t° and E_1 in (54) can be unified into one term since both depend on $\psi_t'^2$

$$E_t^\circ + E_1 = Ga\tilde{I}_t \psi_t'^2 \tag{57}$$

where

$$\tilde{I}_t = (1+C) I_t^\circ, \quad C = \frac{E_1}{E_t^\circ} \tag{58a, b}$$

$\tilde{I}_t$ is the effective torsional modulus which includes both open cross-section and deck effects.

Engine room structure is designed in such a way that the hold double skin continuity is ensured and necessary decks are inserted between the double skins. Strain energy is derived for the first (main) deck and for the others it can be assumed that their strain energy is proportional to the deck plating volume, V, and linearly increasing deformation with the deck distance from inner bottom, h, Fig. 9, since the double bottom is much stiffer than decks. In that way the coefficient C, Eq. (58b), by employing (53) and (56), reads

$$C = \frac{\sum E_i}{E_t} = \frac{4(1+\nu)t_1\left(\frac{a}{b}\right)^3\left(|w_D|+|w_B|\right)^2 k}{\left[1+2(1+\nu)\left(\frac{a}{b}\right)^2\right]I_t^{\circ}a} \tag{59}$$

where

$$k = \sum \frac{V_i}{V_1}\left(\frac{h_i}{h_1}\right)^2. \tag{60}$$

In the above consideration distortion of cross-sections is not included and that is subject of further investigation.

7.2 Torsion of segmented girder

Let us consider a girder consisted of three segments, Fig. 10. The end segments are open and the middle one is closed, so that the girder is symmetric with respect to the z axis. Each segment is specified in its local coordinate system. The properties of the middle and end segments are designated by $(.)^*$ and $(.)^{\circ}$, respectively. The relevant expressions for displacements and sectional forces are listed below (Senjanović et al., 2009a, 2010a):

$$\begin{aligned}
u &= w\frac{\mathrm{d}\psi_t}{\mathrm{d}x} = w\left(\frac{A_1}{l} + A_2 a\,\mathrm{sh}ax + A_3 a\,\mathrm{ch}ax + \psi_p'\right),\\
T_t &= GI_t\left(\frac{A_1}{l} + A_2 a\,\mathrm{sh}ax + A_3 a\,\mathrm{ch}ax + \psi_p'\right),\\
T_w &= -GI_t\left(A_2 a\,\mathrm{sh}ax + A_3 a\,\mathrm{ch}ax\right) - EI_w\psi_p''',\\
T &= GI_t\left(\frac{A_1}{l} + \psi_p'\right) - EI_w\psi_p''',
\end{aligned} \tag{61}$$

$$B_w = GI_t\left(A_2\mathrm{ch}ax + A_3\mathrm{sh}ax\right) + EI_w\psi_p'',$$

where ψ_p represents particular solution of differential equation and coefficient α yields

$$a = \sqrt{\frac{GI_t}{EI_w}}. \tag{62}$$

The symbols A_i and B_i are used for the integration constants of the closed and open segments. The girder is loaded with torque M_t at the ends, while $\mu_x = 0$. The ends are fixed against warping.

The boundary and compatibility conditions in the considered case, yield

$$\psi^{*}(a)=\psi^{\circ}(0),\ \psi_t^{*\prime}(a)=\psi_t^{\circ\prime}(0),$$
$$T^{*}(a)=T^{\circ}(0),\ B_w^{*}(a)=B_w^{\circ}(0), \tag{63}$$
$$u^{\circ}(l^{\circ})=0,\ T^{\circ}(l^{\circ})=M_t.$$

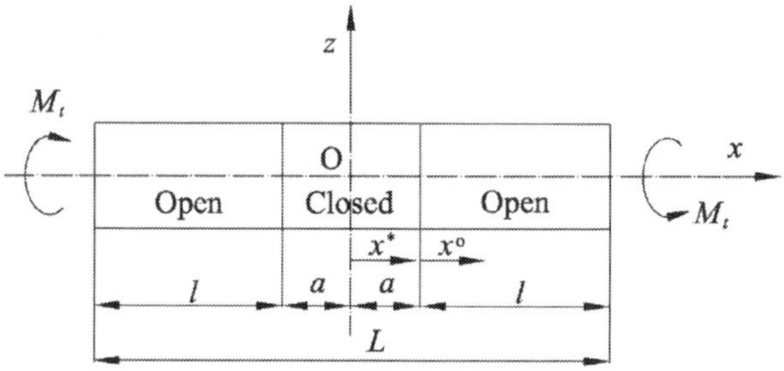

Fig. 10. Torsion of segmented girder

From the third and last conditions (63) one finds

$$A_1=\frac{M_t a}{G\tilde{I}_t},\ B_1=\frac{M_t l^{\circ}}{GI_t^{\circ}}. \tag{64}$$

The remaining four conditions (63) lead to the system of algebraic equations (Senjanović et al., 2010a) and its analytical solution reads:

$$A_3=\frac{D_{A3}}{D},\ B_2=\frac{D_{B2}}{D},\ B_3=\frac{D_{B3}}{D}, \tag{65}$$

where

$$D_{A3}=\frac{M_t}{G}\left[\left(1-\frac{I_t^{\circ}}{\tilde{I}_t}\right)\mathrm{ch}\beta l^{\circ}-1\right],$$
$$D_{B2}=-\frac{M_t}{G}\mathrm{sh}\alpha a\left[\left(1-\frac{\tilde{I}_t}{I_t^{\circ}}\right)\mathrm{ch}\beta l^{\circ}+\frac{\tilde{I}_t}{I_t^{\circ}}\right], \tag{66}$$
$$D_{B3}=\frac{M_t}{G}\left[\left(1-\frac{\tilde{I}_t}{I_t^{\circ}}\right)\mathrm{sh}\alpha a\ \mathrm{sh}\beta l^{\circ}-\frac{a}{\beta}\mathrm{ch}\alpha a\right],$$
$$D=I_t^{\circ}a\mathrm{ch}\alpha a\ \mathrm{ch}\beta l^{\circ}+\tilde{I}_t\beta\mathrm{sh}\alpha a\ \mathrm{sh}\beta l^{\circ}.$$

8. Numerical procedure for vibration analysis

A thin-walled girder is modelled with a set of beam finite elements. Their assemblage in the global coordinate system, performed in the standard way, results in the matrix equation of motion, which may be extended by the damping forces

$$[K]\{\Delta\}+[C]\{\dot{\Delta}\}+[M]\{\ddot{\Delta}\}=\{F(t)\}, \tag{67}$$

where $[K]$, $[C]$ and $[M]$ are the stiffness, damping and mass matrices, respectively; $\{\Delta\},\{\dot{\Delta}\}$ and $\{\ddot{\Delta}\}$ are the displacement, velocity and acceleration vectors, respectively; and $\{F(t)\}$ is the load vector.

In case of natural vibration $\{F(t)\} = \{0\}$ and the influence of damping is rather low for the most of the structures, so that the damping forces may be ignored. Assuming

$$\{\Delta\}=\{\varphi\}\mathrm{e}^{i\omega t}, \tag{68}$$

where $\{\varphi\}$ and ω are the mode vector and natural frequency respectively, Eq. (67) leads to the eigenvalue problem

$$([K]-\omega^2[M])\{\varphi\}=\{0\}, \tag{69}$$

which may be solved by employing different numerical methods (Bathe, 1996) The basic one is the determinant search method in which ω is found from the condition

$$\left|[K]-\omega^2[M]\right|=0 \tag{70}$$

by an iteration procedure. Afterwards, $\{\varphi\}$ follows from (69) assuming unit value for one element in $\{\varphi\}$.

The forced vibration analysis may be performed by direct integration of Eq. (67), as well as by the modal superposition method. In the latter case the displacement vector is presented in the form

$$\{\Delta\}=[\varphi]\{X\}, \tag{71}$$

where $[\varphi]=[\{\varphi\}]$ is the undamped mode matrix and $\{X\}$ is the generalised displacement vector. Substituting (71) into (67), the modal equation yields

$$[k]\{X\}+[c]\{\dot{X}\}+[m]\{\ddot{X}\}=\{f(t)\}, \tag{72}$$

where

$$\begin{aligned}
&[k]=[\varphi]^T[K][\varphi] \text{ - modal stiffness matrix}\\
&[c]=[\varphi]^T[C][\varphi] \text{ - modal damping matrix}\\
&[m]=[\varphi]^T[M][\varphi] \text{ - modal mass matrix}\\
&\{f(t)\}=[\varphi]^T\{F(t)\} \text{ - modal load vector.}
\end{aligned} \tag{73}$$

The matrices $[k]$ and $[m]$ are diagonal, while $[c]$ becomes diagonal only in a special case, for instance if $[C] = \alpha_0 [M] + \beta_0 [K]$, where α_0 and β_0 are coefficients (Senjanović, 1990).

Solving (72) for undamped natural vibration, $[k] = [\omega^2 m]$ is obtained, and by its backward substitution into (72) the final form of the modal equation yields

$$\left[\omega^2\right]\{X\}+2[\omega][\zeta]\{\dot{X}\}+\{\ddot{X}\}=\{\varphi(t)\}, \tag{74}$$

where

$$[\omega]=\left[\sqrt{\frac{k_{ii}}{m_{ii}}}\right] \text{ - natural frequency matrix}$$

$$[\zeta]=\left[\frac{c_{ij}}{2\sqrt{(k_{ii}m_{ii})}}\right] \text{ - relative damping matrix} \tag{75}$$

$$\{\varphi(t)\}=\left\{\frac{f_i(t)}{m_{ii}}\right\} \text{ - relative load vector.}$$

If [ζ] is diagonal, the matrix Eq. (74) is split into a set of uncoupled modal equations.
If vibration excitation is of periodical nature it can be split into harmonics, and the structure response for each of them is determined in the frequency domain. In a case of general or impulsive excitation the vibration problem has to be solved in the time domain.
Several numerical methods are available for this purpose, as for instance the Houbolt, the Newmark and the Wilson θ method (Bathe, 1996), as well as the harmonic acceleration method (Lozina, 1988, Senjanović, 1984).
It is important to point out that all stiffness and mass matrices of the beam finite element (and consequently those of the assembly) are frequency dependent quantities, due to coefficients a and η in the formulation of the shape functions, Eqs. (34) and (35). Therefore, for solving the eigenvalue problem (69) an iteration procedure has to be applied. As a result of frequency dependent matrices, the eigenvectors are not orthogonal. If they are used in the modal superposition method for determining forced response, full modal stiffness and mass matrices are generated. Since the inertia terms are much smaller than the deformation ones in Eqs. (24) and (25), the off-diagonal elements in modal stiffness and mass matrices are very small compared to the diagonal elements and can be neglected.
It is obvious that the usage of the physically consistent non-orthogonal natural modes in the modal superposition method is not practical, especially not in the case of time integration. Therefore, it is preferable to use mathematical orthogonal modes for that purpose. They are created by the static displacement relations yielding from Eqs. (24) and (25) with $\omega=0$, that leads to $a=\eta=1$. In that case all finite element matrices, defined with Eqs. (37) and in Appendix A, can be transformed into explicit form, Appendix B.

9. Cross-section properties of thin-walled girder

Geometrical properties of a thin-walled girder include cross-section area A, moment of inertia of cross-section I_b, shear area A_s, torsional modulus I_t, warping modulus I_w and shear inertia modulus I_s. These parameters are determined analytically for a simple cross-section as pure geometrical properties (Haslum & Tonnessen, 1972, Pavazza, 1991, 2005, Vlasov, 1961).
However, determination of cross-section properties for an open multi-cell cross-section, as for instance in case of ship structures, is quite a difficult task. Therefore, the strip element method is applied for solving this statically indetermined problem (Cheung, 1976). That is well-known and widely used theory of thin-walled girders, which is only briefly described

here. Firstly, axial node displacements are calculated due to bending caused by shear force, and due to torsion caused by variation of twist angle. Then, shear stress in bending τ_b, shear stress due to pure torsion τ_t, shear and normal stresses due to restrained warping τ_w and σ_w, respectively, are determined. Based on the equivalence of strain energies induced by sectional forces and calculated stresses, it is possible to specify cross-section properties in the same formulation as presented below. Furthermore, those formulae can be expressed by stress flows, i.e. stresses due to unit sectional forces (Senjanović & Fan, 1992, 1993).
Shear area:

$$A_s = \frac{Q^2}{\int\limits_A \tau_b^2 dA} = \frac{1}{\int\limits_A g_b^2 dA}, \; g_b = \frac{\tau_b}{Q} . \tag{76}$$

Torsional modulus:

$$I_t = \frac{T_t^2}{\int\limits_A \tau_t^2 dA} = \frac{1}{\int\limits_A g_t^2 dA}, \; g_t = \frac{\tau_t}{T_t} . \tag{77}$$

Shear inertia modulus:

$$I_s = \frac{T_w^2}{\int\limits_A \tau_w^2 dA} = \frac{1}{\int\limits_A g_w^2 dA}, \; g_w = \frac{\tau_w}{T_w} . \tag{78}$$

Warping modulus:

$$I_w = \frac{B_w^2}{\int\limits_A \sigma_w^2 dA} = \frac{1}{\int\limits_A f_w^2 dA}, \; f_w = \frac{\sigma_w}{B_w}; \quad I_w = \int\limits_A w^2 dA . \tag{79}$$

The above quantities are not pure geometrical cross-section properties any more, since they also depend on Poisson's ratio as a physical parameter.
The mass parameters can be expressed with the given mass distribution per unit length, m, and calculated cross-section parameters, i.e.

$$J_b = \frac{m}{A} I_b, \; J_t^0 = \frac{m}{A} I_p, \; J_w = \frac{m}{A} I_w . \tag{80}$$

where $I_p = I_{by} + I_{bz}$ is the polar moment of inertia of cross-section.

10. Illustrative numerical examples

For the illustration of the procedure related to engine room effective stiffness determination, 3D FEM analysis of ship-like pontoon has been undertaken. The 3D FEM model is constituted according to 7800 TEU container ship with main dimensions $L_{pp}xBxH = 319x42.8x24.6$ m, and detailed desciption given in (Tomašević, 2007). The complete hydroelastic analysis of the same ship has been performed.
Stiffness properties of ship hull are calculated by program STIFF, based on the theory of thin-walled girders (STIFF, 1990), Fig. 11.

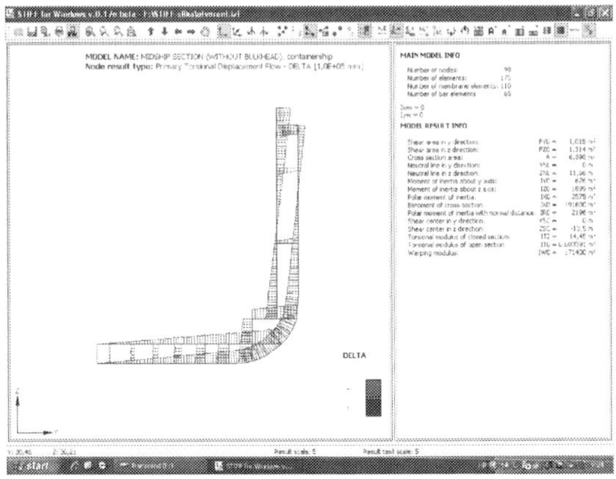

Fig. 11. Program STIFF - warping of ship cross-section

Influence of the transverse bulkheads is taken into account by using the equivalent torsional modulus for the open cross-sections instead of the actual values, i.e. $I_t^* = 2.4 I_t$. This value is applied for all ship-cross sections as the first approximation.

10.1 Analysis of ship-like segmented pontoon

Torsion of the segmented pontoon of the length $L = 300$ m, with effective parameters is considered. Torsional moment $M_t = 40570$ kNm is imposed at the pontoon ends. The pontoon is considered free in the space and the problem is solved analytically according to the formulae given in Section 4. The following values of the basic parameters are used: $a = 10.1$ m, $b = 19.17$ m, $t_1 = 0.01645$ m, $w_D = -221$ m², $w_B = 267$ m², $I_t^\circ = 14.45$ m⁴, $k = 1.894$. As a result $C = 22.42$, Eq. (59), and accordingly $\tilde{I}_t = 338.4$ m⁴, Eq. (58a), are obtained. Since $\tilde{I}_t = 0.36 I_t^*$, effect of the short engine room structure on its torsional stiffness is obvious.

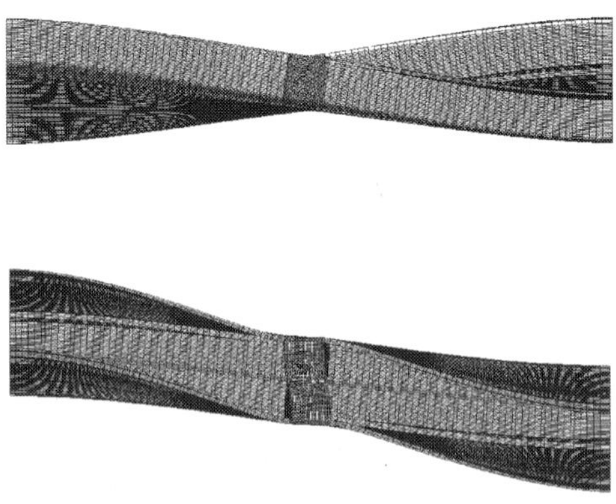

Fig. 12. Deformation of segmented pontoon, lateral and bird view

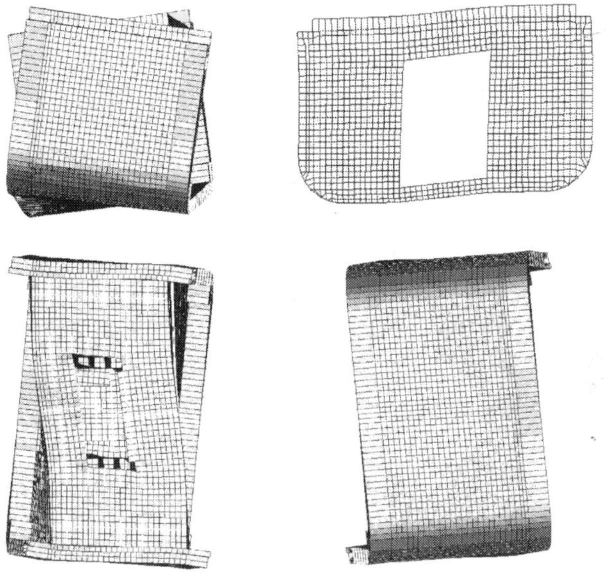

Fig. 13. Lateral, axial, bird and fish views on deformed engine room superelement

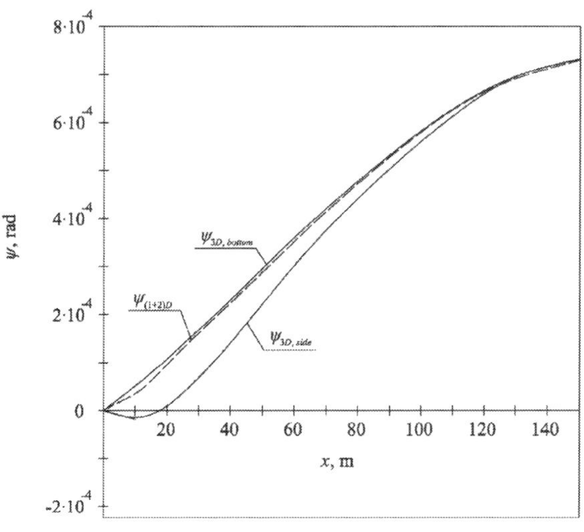

Fig. 14. Twist angles of segmented pontoon

The 3D FEM model of segmented pontoon is made by commercial software package SESAM and consists of 20 open and 1 closed (engine room) superelement. The pontoon ends are closed with transverse bulkheads. The shell finite elements are used. The pontoons are loaded at their ends with the vertical distributed forces in the opposite directions, generating total torque M_t = 40570 kNm. The midship section is fixed against transverse and vertical displacements, and the pontoon ends are constrained against axial displacements (warping). Lateral and bird view on the deformed segmented pontoon is shown in Fig. 12, where the influence of more rigid engine room structure is evident. Detailed view on this pontoon portion is presented in Fig. 13. It is apparent that segment of very stiff double bottom and sides rotate as a "rigid body", while decks and transverse bulkheads are exposed to shear deformation. This deformation causes the distortion of the cross-section, Fig. 13.

Twist angles of the analytical beam solution and that of 3D FEM analysis for the pontoon bottom are compared in Fig. 14. As it can be noticed, there are some small discrepancies between $\psi_{(1+2)D}$ and $\psi_{3D,bottom}$, which are reduced to a negligible value at the pontoon ends

Fig. 14 also shows twist angle of side structure and the difference $\delta = \psi_{3D,\text{bottom}} - \psi_{3D,\text{side}}$ represents distortion angle of cross-section which is highly pronounced. As it is mentioned before, the problem will be further investigated.

10.2 Validation of 1D FEM model

The reliability of 1D FEM analysis is verified by 3D FEM analysis of the considered ship. For this purpose, the light weight loading condition of dry ship with displacement Δ=33692 t is taken into account. The equivalent torsional stiffness of the engine room structure, as well as equivalent stiffness of fore and aft peaks is not taken into account in this example for the time being. However, it will be done in the next step of investigation. The lateral and bird view of the first dominantly torsional and second dominantly horizontal mode of the wetted surface, determined by 1D model, is shown in Fig. 15.

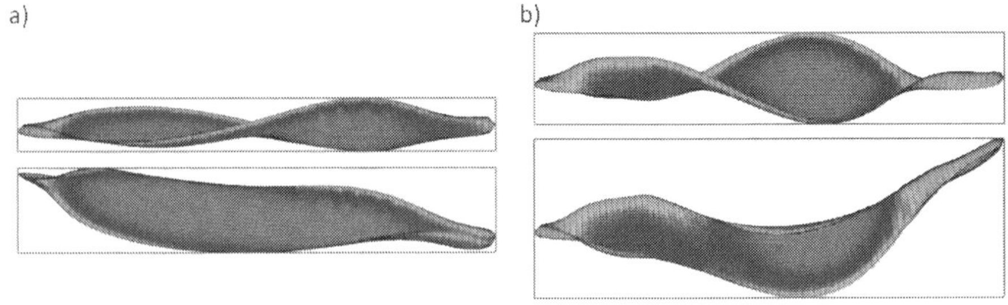

Fig. 15. The first and second mode, lateral and bird view, light weight, 1D model

The first and second 3D dry coupled natural modes of the complete ship structure are shown in Fig. 16. They are similar to that of 1D analysis for the wetted surface. Warping of the transverse bulkheads, which increases the hull torsional stiffness, is evident.

The first four corresponding natural frequencies obtained by 1D and 3D analyses are compared in Table 1.

Mode no.	Vert.		Horiz. + tors.		Mode no.
	1D	3D	1D	3D	
1	7.35	7.33	4.17	4.15	1(H0 + T1)
2	15.00	14.95	7.34	7.40	2(H1 + T2)
3	24.04	22.99	12.22	12.09	3(H2 + T3)
4	35.08	34.21	15.02	16.22	4(H3 + T4)

Table 1. Dry natural frequencies, light weight, ω_i [rad/s]

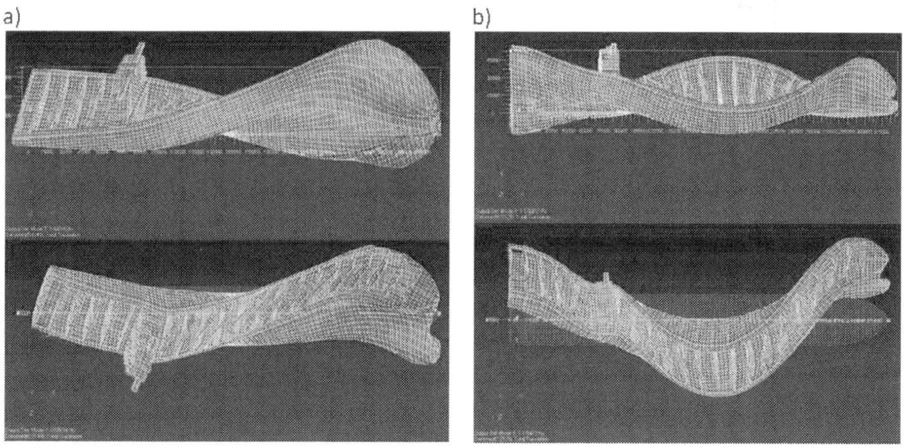

Fig. 16. The first and second mode, lateral and bird view, light weight, 3D model

Quite good agreement is achieved. Values of natural frequencies for higher modes are more difficult to correlate, since strong coupling between global hull modes and local substructure modes of 3D analysis occurs.

10.3 Hydroelastic response of large container ship

Transfer functions of torsional moment and horizontal bending moment at the midship section, obtained using 1D structural model, are shown in Figs. 17 and 18, respectively. The angle of 180° is related to head sea. They are compared to the rigid body ones determined by program HYDROSTAR. Very good agreement is obtained in the lower frequency domain, where the ship behaves as a rigid body, while large discrepancies occur at the resonances of the elastic modes, as expected.

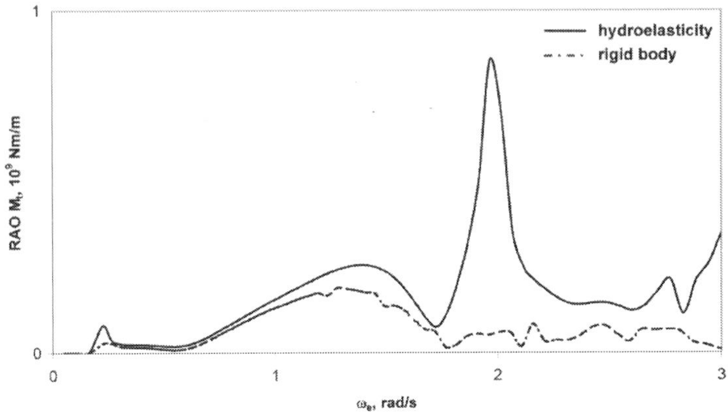

Fig. 17. Transfer function of torsional moment, χ=120°, U=25 kn, x=155.75 m from AP

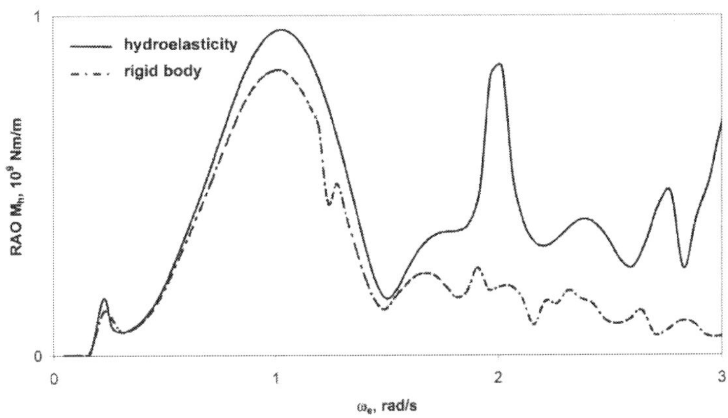

Fig. 18. Transfer function of horizontal bending moment, χ=120°, U=25 kn, x=155.75 m from AP

11. Conclusion

Ultra large container ships are quite elastic and especially sensitive to torsion due to large deck openings. The wave induced response of such ships should be determined by using mathematical hydroelastic models which are consisted of structural, hydrostatic and hydrodynamic parts.

In this chapter the methodology of ship hydroelastic analysis is briefly described, and the role of structural model is discussed. After that, full detail description of the sophisticated beam structural model, which takes shear influence on torsion, as well as contribution of transverse bulkheads and engine room structure to the hull stiffness, is given. Numerical procedure for vibration analysis is also described and determination of ship cross-section properties is explained. The developed theories are illustrated through the numerical examples which include analysis of torsional response of a ship-like segmented pontoon, free vibration analysis of a large container ship and comparison with the results obtained using 3D FEM model, and complete global hydroelastic analysis of a container ship.

It is shown that the used sophisticated beam model of ship hull, based on the advanced thin-walled girder theory with included shear influence on torsion and a proper contribution of transverse bulkheads and engine room structure to its stiffness, is a reasonable choice for determining wave load effects. However, based on the experience, stress concentration in hatch corners calculated directly by the beam model is underestimated. This problem can be overcome by applying substructure approach, i.e. 3D FEM model of substructure with imposed boundary conditions from beam response. In any case, 3D FEM model of complete ship is preferable from the viewpoint of determining stress concentration. Concerning further improvements of the beam model, the distortion induced by torsion is of interest.

The illustrative numerical example of the 7800 TEU container ship shows that the developed hydroelasticity theory, utilizing sophisticated 1D FEM structural model and 3D hydrodynamic model, is an efficient tool for application in ship hydroelastic analyses. The obtained results point out that the transfer functions of hull sectional forces in case of resonant vibration (springing) are much higher than in resonant ship motion.

12. Acknowledgment

This investigation is carried out within the EU FP7 Project TULCS (*Tools for Ultra Large Container Ships*) and the project of Croatian Ministry of Science, Education and Sports *Load and Response of Ship Structures*.

13. Appendix A – consistent finite element properties (frequency dependent formulation)

The stiffness and mass matrices, Eqs. (37), are expressed with one or two integrals, which can be classified in three different types. For general notation of shape functions

$$g_i = \langle g_{ik} \rangle \{ \xi^k \}, \quad i = 1,2,3,4; \quad k = 0,1,2,3, \tag{A1}$$

where q_{ik} are coefficients and $\xi = x / l$, one finds the solutions of integrals in the following form:

$$\begin{aligned} I_{ij}^{(0)}\left(g_{ik}, g_{jk}\right) &= \int_0^l g_i g_j \mathrm{d}x = \langle g_{ik} \rangle \int_0^l \{\xi^k\}\langle \xi^k \rangle \mathrm{d}x \{g_{jk}\} \\ &= l\Big[g_{i0}g_{j0} + \frac{1}{2}\left(g_{i1}g_{j0} + g_{i0}g_{j1}\right) + \frac{1}{3}\left(g_{i2}g_{j0} + g_{i1}g_{j1} + g_{i0}g_{j2}\right) \\ &+ \frac{1}{4}\left(g_{i0}g_{j3} + g_{i1}g_{j2} + g_{i2}g_{j1} + g_{i3}g_{j0}\right) + \frac{1}{5}\left(g_{i1}g_{j3} + g_{i2}g_{j2} + g_{i3}g_{j1}\right) \\ &+ \frac{1}{6}\left(g_{i2}g_{j3} + g_{i3}g_{j2}\right) + \frac{1}{7}g_{i3}g_{j3} \Big] \end{aligned} \tag{A2}$$

$$I_{ij}^{(1)}\left(g_{ik},g_{jk}\right)=\int_0^l \frac{\mathrm{d}g_i}{\mathrm{d}x}\frac{\mathrm{d}g_j}{\mathrm{d}x}\mathrm{d}x=\left\langle g_{ik}\right\rangle\int_0^l\left\{k\xi^{k-1}\right\}\left\langle k\xi^{k-1}\right\rangle\mathrm{d}x\left\{g_{jk}\right\}$$
$$=\frac{1}{l}\Big[g_{i1}g_{j1}+g_{i1}g_{j2}+g_{i2}g_{j1}+g_{i1}g_{j3}+g_{i3}g_{j1}$$
$$+\frac{4}{3}g_{i2}g_{j2}+\frac{3}{2}\left(g_{i2}g_{j3}+g_{i3}g_{j2}\right)+\frac{9}{5}g_{i3}g_{j3}\Big] \tag{A3}$$

$$I_{ij}^{(2)}\left(g_{ik},g_{jk}\right)=\int_0^l \frac{\mathrm{d}^2g_i}{\mathrm{d}x^2}\frac{\mathrm{d}^2g_j}{\mathrm{d}x^2}\mathrm{d}x=\left\langle g_{ik}\right\rangle\int_0^l\left\{k(k-1)\xi^{k-2}\right\}\left\langle k(k-1)\xi^{k-2}\right\rangle\mathrm{d}x\left\{g_{jk}\right\}$$
$$=\frac{4}{l^3}\left[g_{i2}g_{j2}+\frac{3}{2}\left(g_{i2}g_{j3}+g_{i3}g_{j2}\right)+3g_{i3}g_{j3}\right]. \tag{A4}$$

Thus, the finite element properties can be written in the following systematic way suitable for coding.

Stiffness matrices

$$\begin{aligned}[k]_{bs}&=EI_b\left[I_{ij}^{(2)}\left(a_{ik},a_{jk}\right)\right]+GA_s\left[I_{ij}^{(1)}\left(b_{ik},b_{jk}\right)\right]\\ [k]_{ws}&=EI_w\left[I_{ij}^{(2)}\left(d_{ik},d_{jk}\right)\right]+GI_s\left[I_{ij}^{(1)}\left(e_{ik},e_{jk}\right)\right]\\ [k]_t&=GI_t\left[I_{ij}^{(1)}\left(d_{ik},d_{jk}\right)\right]\end{aligned} \tag{A5}$$

Mass matrices

$$\begin{aligned}[m]_{sb}&=m\left[I_{ij}^{(0)}\left(c_{ik},c_{jk}\right)\right]+J_b\left[I_{ij}^{(1)}\left(a_{ik},a_{jk}\right)\right]\\ [m]_{tw}&=J_t\left[I_{ij}^{(0)}\left(f_{ik},f_{jk}\right)\right]+J_w\left[I_{ij}^{(1)}\left(d_{ik},d_{jk}\right)\right]\\ [m]_{st}&=m_c\left[I_{ij}^{(0)}\left(c_{ik},f_{jk}\right)\right],\ [m]_{ts}=[m]_{st}^T\end{aligned} \tag{A6}$$

Load vectors

$$\begin{aligned}\{q\}&=l\left\{q_0\left(c_{i0}+\frac{1}{2}c_{i1}+\frac{1}{3}c_{i2}+\frac{1}{4}c_{i3}\right)+q_1\left(\frac{1}{2}c_{i0}+\frac{1}{3}c_{i1}+\frac{1}{4}c_{i2}+\frac{1}{5}c_{i3}\right)\right\}\\ \{\mu\}&=l\left\{\mu_0\left(f_{i0}+\frac{1}{2}f_{i1}+\frac{1}{3}f_{i2}+\frac{1}{4}f_{i3}\right)+\mu_1\left(\frac{1}{2}f_{i0}+\frac{1}{3}f_{i1}+\frac{1}{4}f_{i2}+\frac{1}{5}f_{i3}\right)\right\}\end{aligned} \tag{A7}$$

14. Appendix B – simplified finite element properties, from appendix A (frequency independent formulation)

Stiffness matrices:

$$[k]_{bs}=\frac{2EI_b}{(1+12\beta)l^3}\begin{bmatrix}6 & 3l & -6 & 3l\\ & 2(1+3\beta)l^2 & -3l & (1-6\beta)l^2\\ & & 6 & -3l\\ Sym. & & & 2(1+3\beta)l^2\end{bmatrix} \tag{B1}$$

$$\left[k\right]_{ws}=\frac{2EI_w}{(1+12\gamma)l^3}\begin{bmatrix} 6 & 3l & -6 & 3l \\ & 2(1+3\gamma)l^2 & -3l & (1-6\gamma)l^2 \\ & & 6 & -3l \\ Sym. & & & 2(1+3\gamma)l^2 \end{bmatrix} \tag{B2}$$

$$\left[k\right]_{t}=\frac{GI_t}{30(1+12\gamma)^2 l}\begin{bmatrix} 36 & 3(1-60\gamma)l & -36 & 3(1-60\gamma)l \\ & 4(1+15\gamma+360\gamma^2)l^2 & -3(1-60\gamma)l & -(1+60\gamma-720\gamma^2)l^2 \\ & & 36 & -3(1-60\gamma)l \\ Sym. & & & 4(1+15\gamma+360\gamma^2)l^2 \end{bmatrix} \tag{B3}$$

Mass matrices:

$$\left[m\right]_{sb}=\left[m\right]_{s}+\left[m\right]_{b} \tag{B4}$$

$$\left[m\right]_{s}=\frac{ml}{420(1+12\beta)^2}\begin{bmatrix} 156+3528\beta+20160\beta^2 & (22+462\beta+2520\beta^2)l & 54+1512\beta+10080\beta^2 & -(13+378\beta+2520\beta^2)l \\ & (4+84\beta+504\beta^2)l^2 & (13+378\beta+2520\beta^2)l & -(3+84\beta+504\beta^2)l^2 \\ & & 156+3528\beta+20160\beta^2 & -(22+462\beta+2520\beta^2)l \\ Sym. & & & (4+84\beta+504\beta^2)l^2 \end{bmatrix} \tag{B5}$$

$$\left[m\right]_{b}=\frac{J_b}{30(1+12\beta)^2 l}\begin{bmatrix} 36 & (3-180\beta)l & -36 & (3-180\beta)l \\ & (4+60\beta+1440\beta^2)l^2 & (-3+180\beta)l & -(1+60\beta-720\beta^2)l^2 \\ & & 36 & (-3+180\beta)l \\ Sym. & & & (4+60\beta+1440\beta^2)l^2 \end{bmatrix} \tag{B6}$$

$$\left[m\right]_{tw}=\left[m\right]_{t}+\left[m\right]_{w} \tag{B7}$$

$$\left[m\right]_{t}=\frac{J_t l}{420(1+12\gamma)^2}\begin{bmatrix} 156+3528\gamma+20160\gamma^2 & (22+462\gamma+2520\gamma^2)l & 54+1512\gamma+10080\gamma^2 & -(13+378\gamma+2520\gamma^2)l \\ & (4+84\gamma+504\gamma^2)l^2 & (13+378\gamma+2520\gamma^2)l & -(3+84\gamma+504\gamma^2)l^2 \\ & & 156+3528\gamma+20160\gamma^2 & -(22+462\gamma+2520\gamma^2)l \\ Sym. & & & (4+84\gamma+504\gamma^2)l^2 \end{bmatrix} \tag{B8}$$

$$\left[m\right]_{w}=\frac{J_w}{30(1+12\gamma)^2 l}\begin{bmatrix} 36 & (3-180\gamma)l & -36 & (3-180\gamma)l \\ & (4+60\gamma+1440\gamma^2)l^2 & (-3+180\gamma)l & -(1+60\gamma-720\gamma^2)l^2 \\ & & 36 & (-3+180\gamma)l \\ Sym. & & & (4+60\gamma+1440\gamma^2)l^2 \end{bmatrix} \tag{B9}$$

$$[m]_{st} = \frac{mlc}{420(1+12\beta)(1+12\gamma)} \times \begin{bmatrix} 156+1764\beta+1764\gamma+20160\beta\gamma & (22+252\beta+210\gamma+2520\beta\gamma)l & 54+756\beta+756\gamma+10080\beta\gamma & -(13+168\beta+210\gamma+2520\beta\gamma)l \\ (22+210\beta+252\gamma+2520\beta\gamma)l & (4+42\beta+42\gamma+504\beta\gamma)l^2 & (13+210\beta+168\gamma+2520\beta\gamma)l & -(3+42\beta+42\gamma+504\beta\gamma)l^2 \\ 54+756\beta+756\gamma+10080\beta\gamma & (13+168\beta+210\gamma+2520\beta\gamma)l & 156+1764\beta+1764\gamma+20160\beta\gamma & -(22+252\beta+210\gamma+2520\beta\gamma)l \\ -(13+210\beta+168\gamma+2520\beta\gamma)l & -(3+42\beta+42\gamma+504\beta\gamma)l^2 & -(22+210\beta+252\gamma+2520\beta\gamma)l & (4+42\beta+42\gamma+504\beta\gamma)l^2 \end{bmatrix} \quad \text{(B10)}$$

$$[m]_{ts} = [m]_{st}^T \quad \text{(B11)}$$

Load vectors:

$$\{q\} = \frac{q_0 l}{12} \begin{Bmatrix} 6 \\ l \\ 6 \\ -l \end{Bmatrix} + \frac{q_1 l}{60(1+12\beta)} \begin{Bmatrix} 9+120\beta \\ (2+30\beta)l \\ 21+240\beta \\ -(3+30\beta)l \end{Bmatrix} \quad \text{(B12)}$$

$$\{\mu\} = \frac{\mu_0 l}{12} \begin{Bmatrix} 6 \\ l \\ 6 \\ -l \end{Bmatrix} + \frac{\mu_1 l}{60(1+12\gamma)} \begin{Bmatrix} 9+120\gamma \\ (2+30\gamma)l \\ 21+240\gamma \\ -(3+30\gamma)l \end{Bmatrix} \quad \text{(B13)}$$

Stiffness ratios:

$$\beta = \frac{EI_b}{GA_s l^2}, \quad \gamma = \frac{EI_w}{GI_s l^2}. \quad \text{(B14)}$$

15. References

Bathe, KJ. (1996). *Finite Element Procedures*, Prentice Hall

Cheung, YK. (1976). *Finite Strip Method in Structural Analysis*, Pergamon Press

Haslum, K. & Tonnessen, A. (1972). An Analysis of Torsion in Ship Hull, *European Shipbuilding*, No.5/6, pp. 67-89

Kawai, T. (1973). The Application of Finite Element Method to Ship Structures, *Computers & Structures*, Vol.3, No.5, pp. 1175-1194, ISSN 0045-7949

Lozina, Ž. (1988). A Comparison of Harmonic Acceleration Method with the Other Commonly Used Methods for Calculation of Dynamic Transient Response, *Computers & Structures*, Vol.29, No.2, pp. 227-240, ISSN 0045-7949

Pavazza, R. (1991). *Bending and Torsion of Thin-Walled Beams of Open Section on Elastic Foundation*, Ph.D. Thesis. University of Zagreb, (in Croatian)

Pavazza, R. (2005). Torsion of Thin-Walled Beams of Open Cross-Sections with Influence of Shear, *International Journal of Mechanical Sciences*, Vol.47, No.7, pp. 1099-1122, ISSN 0020-7403

Pedersen, PT. (1983). A Beam Model for the Torsional-Bending Response of Ships Hulls, *RINA Transactions*, Vol.31, pp. 171-182

Pedersen, PT. (1985). Torsional Response of Container Ships, *Journal of Ship Research,* Vol.29, pp. 194-205, ISSN 1542-0604

Senjanović, I. (1984). Harmonic Acceleration Method for Dynamic Structural Analysis, *Computers & Structures,* Vol.18, No.1, pp. 71-80, ISSN 0045-7949

Senjanović, I. (1990). *Ship Vibrations,* Part II, University of Zagreb, (in Croatian)

Senjanović, I. & Fan, Y. (1989). A Higher-Order Flexural Beam Theory, *Computers & Structures,* Vol.32, No.5, pp. 973-986, ISSN 0045-7949

Senjanović, I. & Fan, Y. (1992). A Higher-Order Theory of Thin-Walled Girders with Application to Ship Structures, *Computers & Structures,* Vol.43, No.1, pp. 31-52, ISSN 0045-7949

Senjanović, I. & Fan, Y. (1993). A Finite Element Formulation of Initial Ship Cross-Section Properties, *Brodogradnja,* Vol.41, No.1, pp. 27-36, ISSN 0007-215X

Senjanović, I. & Fan, Y. (1997). A Higher-Order Torsional Beam Theory, *International Journal for Engineering Modelling,* Vol.32, No.1-4, pp. 25-40, ISSN 1330-1365

Senjanović, I. & Grubišić, R. (1991). Coupled Horizontal and Torsional Vibration of a Ship Hull with Large Hatch Openings, *Computers & Structures,* Vol.41, No.2, pp. 213-226, ISSN 0045-7949

Senjanović, I., Malenica, Š., Tomašević, S. & Rudan, S. (2007). Methodology of Ship Hydroelastic Investigation, *Brodogradnja,* Vol.58, No.2, pp. 133-145, ISSN 0007-215X

Senjanović, I. , Tomašević, S., Tomić, M., Rudan, S., Vladimir N. & Malenica, Š. (2008a). Hydroelasticity of Very Large Container Ships, *Proceedings of International Conference on Design and Operation of Container Ships,* pp. 51-70, RINA, London

Senjanović, I., Tomašević, S., Rudan, S. & Senjanović, T. (2008b). Role of Transverse Bulkheads in Hull Stiffness of Large Container Ships, *Engineering Structures,* Vol.30, No.9, pp. 2492-2509, ISSN 0141-0296

Senjanović, I., Tomašević, S. & Vladimir, N. (2009a). An Advanced Theory of Thin-Walled Girders with Application to Ship Vibrations, *Marine Structures,* Vol.22, No.3, pp. 387-437, ISSN 0951-8339

Senjanović, I. , Tomašević, S., Vladimir N. & Malenica, Š. (2009b). Numerical Procedure for Ship Hydroelastic Analysis, *Proceedings of International Conference on Computational Methods in Marine Engineering,* pp. 259-264, CIMNE, Barcelona

Senjanović, I., Vladimir, N. & Tomić, M. (2010a). The Contribution of the Engine Room Structure to the Hull Stiffness of Large Container Ships, *International Shipbuilding Progress,* Vol.57, No.1-2, pp. 65-85, ISSN 0020-868X

Senjanović, I. , Tomašević, S., Vladimir N. Tomić, M. & Malenica, Š. (2010b). Application of an Advanced Beam Theory to Ship Hydroelastic Analysis, *Proceedings of International Workshop on Advanced Ship Design for Pollution Prevention,* pp. 31-42, Taylor & Francis, London

STIFF (1990). *User's Manual,* University of Zagreb

Szilard, R. (2004). *Theories and Applications of Plate Analysis,* John Wiley & Sons, New York

Timoshenko, S. & Young, DH. (1955). *Vibrations Problems in Engineering,* D. Van Nostrand

Tomašević, S. (2007). *Hydroelastic Model of Dynamic Response of Container Ship in Waves,* Ph.D. Thesis. University of Zagreb, (in Croatian)

Vlasov, VZ. (1961). *Thin-Walled Elastic Beams,* Israel Program for Scientific Translation, Jerusalem

Wu, YS. & Ho, CS. (1987). Analysis of Wave Induced Horizontal and Torsion Coupled Vibrations of Ship Hull. *Journal of Ship Research,* Vol.31, No.4, pp. 235-252, ISSN 1542-0604

11

Stochastic Finite Element Method in Mechanical Vibration

Mo Wenhui
Hubei University Of Automotive Technology
China

1. Introduction

Material properties, geometry parameters and applied loads of the structure are assumed to be stochastic. Although the finite element method analysis of complicated structures has become a generally widespread and accepted numerical method in the world, regarding the given factors as constants can not apparently correspond to the reality of a structure.
The direct Monte Carlo simulation of the stochastic finite element method(DSFEM) requires a large number of samples, which requires much calculation time and occupies much computer storage space [1]. Monte Carlo simulation by applying the Neumann expansion (NSFEM) enhances computational efficiency and saves storage in such a way that the NSFEM combined with Monte Carlo simulation enhances the finite element model advantageously [2]. The preconditioned Conjugate Gradient method (PCG) applied in the calculation of stochastic finite elements can also enhance computational accuracy and efficiency [3]. The TSFEM assumes that random variables are dealt with by Taylor expansion around mean values and is obtained by appropriate mathematical treatment [4, 14]. According to first-order or second-order perturbation methods, calculation formulas can be obtained [2, 5, 6,8, 9, 13, 15, 16]. The result is called the PSFEM and has been adopted by many scholars.
The PSFEM is often applied in dynamic analysis of structures and the second- order perturbation technique has been proved to be accurate and efficient. Dynamic reliability of a large frame is calculated by the SFEM and response sensitivity is formulated in the context of stiffness and mass matrix condensation [7]. Nonlinear structural dynamics are developed by the PSFEM. Nonlinearities due to material and geometrical effects have also been included [8]. By forming a new dynamic shape function matrix, dynamic analysis of the spatial frame structure is presented by the PSFEM [9].
It is significant to extend this research to the dynamic state. Considering the influence of random factors, the mechanical vibrations for a linear system are illustrated by using the TSFEM and the CG.

2. Random variable

Material properties, geometry parameters and applied loads of machines are assumed to be independent random variables, and are indicated as $a_1, a_2, \cdots, \ a_i, \cdots, a_{n_1}$. Their means are $\mu_1, \mu_2, \cdots, \mu_i, \cdots, \mu_{n_1}$, and their variances are.. $\sigma_i^{\ 2}, \cdots, \sigma_{n_1}^{\ \ 2}$. When they are subject to normal distributions, the standard method used to simulate them is to take advantage of well-tested computer programs. When they are subject to unknown distributions, the sample of $a_1, a_2, \cdots, a_i, \cdots, a_{n_1}$ can be generated from the following method:

$$P\{|x-\mu'| \geq \varepsilon\} \leq \frac{\sigma^2}{\varepsilon^2} \tag{1}$$

where, x is a random variable, μ' is the mean, σ is the standard deviation ,and ε is an arbitrary positive number. Eq.1 is called the Chebyschev inequality.
The Chebyschev inequality can also be written as

$$P\{|x-\mu'| < \varepsilon\} \geq 1-\frac{\sigma^2}{\varepsilon^2} \tag{2}$$

After substituting $\varepsilon = 6\sigma_i$, $x = a_i$, Eq.2 becomes

$$P\{|a_i-\mu_i| < 6\sigma_i\} \geq 0.9722 \tag{3}$$

where

$$a_i < 6\sigma_i + \mu_i \tag{4}$$

or

$$a_i > -6\sigma_i + \mu_i \tag{5}$$

If it is assumed that z' is a random number within the open interval (0,1), then

$$a_i = 6\sigma_i z' + \mu_i \tag{6}$$

or

$$a_i = -6\sigma_i z' + \mu_i \tag{7}$$

Large numbers of samples of random variables $a_1, a_2, \cdots, a_i, \cdots, a_{n_1}$ are produced from Eqs.6 and 7 so as to resolve the stochastic finite element problem through Monte Carlo stimulation.

3. Dynamic analysis of finite element

For a linear system, the dynamic equilibrium equation is given by

$$[M]\{\ddot{\delta}\}+[C]\{\dot{\delta}\}+[K]\{\delta\}=\{F\} \tag{8}$$

where $\{\ddot{\delta}\},\{\dot{\delta}\},\{\delta\}$ are the acceleration, velocity and displacement vectors. $[M],[K]$ and $[C]$ are the global mass, stiffness and damping matrices obtained by assembling the element variables in global coordinate system.
In order to program easily, the comprehensive calculation steps of the Newmark method are as follows

1. The initial calculation

The matrices $[K]$, $[M]$ and $[C]$ are formed.

The initial values $\{\delta_t\},\{\dot{\delta}_t\},\{\ddot{\delta}_t\}$ are given.

After selecting step Δt and parameters γ,β, the following relevant parameters are calculated:

$$\gamma \geq 0.50 \qquad \beta \geq 0.25(0.5+\gamma)^2$$

$$b_0 = \frac{1}{\beta(\Delta t)^2} \quad b_1 = \frac{\gamma}{\beta \Delta t} \quad b_2 = \frac{1}{\beta \Delta t}$$

$$b_3 = \frac{1}{2\beta} - 1 \quad b_4 = \frac{\gamma}{\beta} - 1 \quad b_5 = \frac{\Delta t}{2}\left(\frac{\gamma}{\beta} - 2\right)$$

$$b_6 = \Delta t(1-\gamma) \qquad b_7 = \gamma \Delta t$$

The stiffness matrix is defined as

$$\left[\tilde{K}\right] = [K] + b_0[M] + b_1[C] \tag{9}$$

The stiffness matrix inversion $\left[\tilde{K}\right]^{-1}$ is solved.

2. Calculation of each step time

At time $t+\Delta t$, the load vector is defined as

$$\begin{aligned}\left\{\tilde{F}_{t+\Delta t}\right\} &= \left\{F_{t+\Delta t}\right\} + [M]\left(b_0\{\delta_t\} + b_2\{\dot{\delta}_t\} + b_3\{\ddot{\delta}_t\}\right) \\ &+ [C]\left(b_1\{\delta_t\} + b_4\{\dot{\delta}_t\} + b_5\{\ddot{\delta}_t\}\right)\end{aligned} \tag{10}$$

At time $t+\Delta t$, the displacement vector is given by

$$\left\{\delta_{t+\Delta t}\right\} = \left[\tilde{K}\right]^{-1}\left\{\tilde{F}_{t+\Delta t}\right\} \tag{11}$$

At time $t+\Delta t$, the velocity vector and acceleration vector are obtained as

$$\left\{\ddot{\delta}_{t+\Delta t}\right\} = b_0\left(\left\{\delta_{t+\Delta t}\right\} - \{\delta_t\}\right) - b_2\left\{\dot{\delta}_t\right\} - b_3\left\{\ddot{\delta}_t\right\} \tag{12}$$

$$\left\{\dot{\delta}_{t+\Delta t}\right\} = \left\{\dot{\delta}_t\right\} + b_6\left\{\ddot{\delta}_t\right\} + b_7\left\{\ddot{\delta}_{t+\Delta t}\right\} \tag{13}$$

Vectors $\left\{\delta_{t+i_1\Delta t}\right\},\left\{\dot{\delta}_{t+i_1\Delta t}\right\},\left\{\ddot{\delta}_{t+i_1\Delta t}\right\}$ are solved at time $t+i_1\Delta t(i_1 = 2,3,\cdots,n_3)$ step-by -step.

4. Analysis of mechanical vibration based on CG

Eq.11 can be expressed as

$$\left[\tilde{K}\right]\left\{\delta_{t+\Delta t}\right\} = \left\{\tilde{F}_{t+\Delta t}\right\} \tag{14}$$

N_1 samples of random variables $a_1, a_2, \cdots, a_i, \cdots, a_{n_1}$ are produced. N_1 matrices $\left[\tilde{K}\right]$ and N_1 Eqs.14 are also generated. For a linear vibration, Eq.14 is the system of linear equations. The CG method is an effective method for solving the large system of linear equations according to the following steps:

1. First, select an approximate solution as the initial value

$$\delta^{(0)}{}_{t+\Delta t} = \left(\delta^{(0)}{}_{(t+\Delta t)_1}, \delta^{(0)}{}_{(t+\Delta t)_2}, \cdots, \delta^{(0)}{}_{(t+\Delta t)_{N''}}\right)^T \tag{15}$$

2. Calculate the first residual vector

$$r^{(0)} = \left\{\tilde{F}_{t+\Delta t}\right\} - \left[\tilde{K}\right]\delta^{(0)}{}_{t+\Delta t} \tag{16}$$

and vector

$$p^{(0)} = \left[\tilde{K}\right]^T r^{(0)} \tag{17}$$

where, $\left[\tilde{K}\right]^T$ is the transposed matrix of $\left[\tilde{K}\right]$

3. For $\bar{i} = 0, 1, 2, \cdots, n_2 - 1$, iterate step-by-step as follows

$$\alpha_{\bar{i}} = \frac{\left(\left[\tilde{K}\right]p^{(\bar{i})}, r^{(\bar{i})}\right)}{\left(\left[\tilde{K}\right]p^{(\bar{i})}, \left[\tilde{K}\right]p^{(\bar{i})}\right)} = \frac{\left(p^{(\bar{i})}, \left[\tilde{K}\right]^T r^{(\bar{i})}\right)}{\left(\left[\tilde{K}\right]p^{(\bar{i})}, \left[\tilde{K}\right]p^{(\bar{i})}\right)} = \frac{\left(\left[\tilde{K}\right]^T r^{(\bar{i})}, \left[\tilde{K}\right]^T r^{(\bar{i})}\right)}{\left(\left[\tilde{K}\right]p^{(\bar{i})}, \left[\tilde{K}\right]p^{(\bar{i})}\right)} \tag{18}$$

$$\left\{\delta_{t+\Delta t}\right\}^{(\bar{i}+1)} = \left\{\delta_{t+\Delta t}\right\}^{(\bar{i})} + \alpha_{\bar{i}} p^{(\bar{i})} \tag{19}$$

$$r^{(\bar{i}+1)} = r^{(\bar{i})} - \alpha_{\bar{i}}\left[\tilde{K}\right]p^{(\bar{i})} \tag{20}$$

$$\beta_{\bar{i}+1} = \frac{\left(\left[\tilde{K}\right]^T r^{(\bar{i}+1)}, \left[\tilde{K}\right]^T r^{(\bar{i}+1)}\right)}{\left(\left[\tilde{K}\right]^T r^{(\bar{i})}, \left[\tilde{K}\right]^T r^{(\bar{i})}\right)} \tag{21}$$

$$p^{(\bar{i}+1)} = \left[\tilde{K}\right]^T r^{(\bar{i}+1)} + \beta_{\bar{i}+1} p^{(\bar{i})} \tag{22}$$

The process can be stopped only if r^{n_2} is small enough.
Vectors $\left\{\delta_{t+\Delta t}\right\}_1, \left\{\delta_{t+\Delta t}\right\}_2, \cdots, \left\{\delta_{t+\Delta t}\right\}_{N_1}$ are solutions of N_1 Eqs.14.
The mean of $\left\{\delta_{t+\Delta t}\right\}$ is given by

$$\mu\left\{\delta_{t+\Delta t}\right\} = \frac{\left\{\delta_{t+\Delta t}\right\}_1 + \left\{\delta_{t+\Delta t}\right\}_2 + \cdots + \left\{\delta_{t+\Delta t}\right\}_{N_1}}{N_1} \tag{23}$$

The variance of $\{\delta_{t+\Delta t}\}$ is given by

$$Var\{\delta_{t+\Delta t}\} = \frac{1}{N_1 - 1}\sum_{i=1}^{N_1}\left(\{\delta_{t+\Delta t}\}_i - \mu\{\delta_{t+\Delta t}\}\right)^2 \tag{24}$$

Similarly, the mean and variance of the vector $\{\delta_{t+i_1\Delta t}\}$ can be solved for at time $t + i_1\Delta t(i_1 = 2,3,\cdots,n_3)$ step-by-step.
At time $t' = t + i_2\Delta t(i_2 = 1,2,\cdots,n_4)$,the strain and stress vectors for element d are

$$\{\varepsilon\} = [B]\{\delta_{t'}{}^{d}\} \tag{25}$$

and

$$\{\sigma\} = [D]\{\varepsilon\} \tag{26}$$

where, $[D]$=the material response matrix of element d ,$[B]$=the gradient matrix of element d and $\{\delta_{t'}{}^{d}\}$ =the element d nodal displacement vector at time t' .
Substituting Eq.25 into Eq.26, the stress for element d is given by

$$\{\sigma\} = [D][B]\{\delta_{t'}{}^{d}\} \tag{27}$$

Substituting..samples of random variables $a_1, a_2,\cdots, a_i,\cdots, a_{n_1}$ into Eq.27, the vectors $\{\sigma\}_1, \{\sigma\}_2,\cdots,\{\sigma\}_{N_1}$ can be obtained.
The mean of $\{\sigma\}$ is given by

$$\mu\{\sigma\} = \frac{\{\sigma\}_1 + \{\sigma\}_2 + \cdots + \{\sigma\}_{N_1}}{N_1} \tag{28}$$

The variance of $\{\sigma\}$ is given by

$$Var\{\sigma\} = \frac{1}{N_1 - 1}\sum_{i=1}^{N_1}\left(\{\sigma\}_i - \mu\{\sigma\}\right)^2 \tag{29}$$

The CG method belongs to method of iteration with the advantage of quick convergence. For practical purpose, PCG is applied to accelerate the convergence.

5. Analysis of mechanical vibration based on TSFEM

Independent random variables of the system are regarded as $a_1, a_2,\cdots, a_i,\cdots, a_{n_1}$.
The partial derivative of Eq.14 with respect to a_i is given by

$$\frac{\partial\{\delta_{t+\Delta t}\}}{\partial a_i} = \left[\tilde{K}\right]^{-1}\left(\frac{\partial\{\tilde{F}_{t+\Delta t}\}}{\partial a_i} - \frac{\partial\left[\tilde{K}\right]}{\partial a_i}\{\delta_{t+\Delta t}\}\right) \tag{30}$$

where

$$\frac{\partial\{\tilde{F}_{t+\Delta t}\}}{\partial a_i}=\frac{\partial\{F_{t+\Delta t}\}}{\partial a_i}+[M]\left(b_0\frac{\partial\{\delta_t\}}{\partial a_i}+b_2\frac{\partial\{\dot{\delta}_t\}}{\partial a_i}+b_3\frac{\partial\{\ddot{\delta}_t\}}{\partial a_i}\right)+$$
$$\frac{\partial[M]}{\partial a_i}\left(b_0\{\delta_t\}+b_2\{\dot{\delta}_t\}+b_3\{\ddot{\delta}_t\}\right)$$
$$+[C]\left(b_1\frac{\partial\{\delta_t\}}{\partial a_i}+b_4\frac{\partial\{\dot{\delta}_t\}}{\partial a_i}+b_5\frac{\partial\{\ddot{\delta}_t\}}{\partial a_i}\right)$$
$$+\frac{\partial[C]}{\partial a_i}\left(b_1\{\delta_t\}+b_4\{\dot{\delta}_t\}+b_5\{\ddot{\delta}_t\}\right) \tag{31}$$

After $\left.\frac{\partial\{\delta_t\}}{\partial a_i}\right|_{t=0}=q_0,\left.\frac{\partial\{\dot{\delta}_t\}}{\partial a_i}\right|_{t=0}=\dot{q}_0,\left.\frac{\partial\{\ddot{\delta}_t\}}{\partial a_i}\right|_{t=0}=\ddot{q}_0$ are given, Eq.31 can be calculated.

The partial derivative of Eq.30 with respect to a_i is given by

$$\frac{\partial^2\{\delta_{t+\Delta t}\}}{\partial a_i^2}=\left[\tilde{K}\right]^{-1}\left(\frac{\partial^2\{\tilde{F}_{t+\Delta t}\}}{\partial a_i^2}-2\frac{\partial\left[\tilde{K}\right]}{\partial a_i}\frac{\partial\{\delta_{t+\Delta t}\}}{\partial a_i}-\frac{\partial^2\left[\tilde{K}\right]}{\partial a_i^2}\{\delta_{t+\Delta t}\}\right) \tag{32}$$

where

$$\frac{\partial^2\{\tilde{F}_{t+\Delta t}\}}{\partial a_i^2}=\frac{\partial^2\{F_{t+\Delta t}\}}{\partial a_i^2}+\frac{\partial^2[M]}{\partial a_i^2}\left(b_0\{\delta_t\}+b_2\{\dot{\delta}_t\}+b_3\{\ddot{\delta}_t\}\right)+$$
$$2\frac{\partial[M]}{\partial a_i}\left(b_0\frac{\partial\{\delta_t\}}{\partial a_i}+b_2\frac{\partial\{\dot{\delta}_t\}}{\partial a_i}+b_3\frac{\partial\{\ddot{\delta}_t\}}{\partial a_i}\right)+[M]\left(b_0\frac{\partial^2\{\delta_t\}}{\partial a_i^2}+b_2\frac{\partial^2\{\dot{\delta}_t\}}{\partial a_i^2}+b_3\frac{\partial^2\{\ddot{\delta}_t\}}{\partial a_i^2}\right)$$
$$+\frac{\partial^2[C]}{\partial a_i^2}\left(b_1\{\delta_t\}+b_4\{\dot{\delta}_t\}+b_5\{\ddot{\delta}_t\}\right)+2\frac{\partial[C]}{\partial a_i}\left(b_1\frac{\partial\{\delta_t\}}{\partial a_i}+b_4\frac{\partial\{\dot{\delta}_t\}}{\partial a_i}+b_5\frac{\partial\{\ddot{\delta}_t\}}{\partial a_i}\right)$$
$$+[C]\left(b_1\frac{\partial^2\{\delta_t\}}{\partial a_i^2}+b_4\frac{\partial^2\{\dot{\delta}_t\}}{\partial a_i^2}+b_5\frac{\partial^2\{\ddot{\delta}_t\}}{\partial a_i^2}\right) \tag{33}$$

After $\left.\frac{\partial^2\{\delta_t\}}{\partial a_i^2}\right|_{t=0}=r_0,\left.\frac{\partial^2\{\dot{\delta}_t\}}{\partial a_i^2}\right|_{t=0}=\dot{r}_0,\left.\frac{\partial^2\{\ddot{\delta}_t\}}{\partial a_i^2}\right|_{t=0}=\ddot{r}_0$ are given, Eq.33 can be calculated.

The displacement is expanded at the mean value point $\bar{a}=\left(\bar{a}_1,\bar{a}_2,\cdots,\bar{a}_i,\cdots,\bar{a}_{n_1}\right)^T$ by means of a Taylor series. By taking the expectation operator for two sides of above Eq.11, the mean of the displacement is obtained as

$$\mu\{\delta_{t+\Delta t}\} \approx \{\delta_{t+\Delta t}\}\Big|_{a=\bar{a}} + \frac{1}{2}\sum_{i=1}^{n_1} \frac{\partial^2\{\delta_{t+\Delta t}\}}{\partial a_i^2}\Bigg|_{a=\bar{a}} \cdot \sigma_i^2 \tag{34}$$

where, $\mu\{\delta_{t+\Delta t}\}$ expresses the mean value of $\delta_{t+\Delta t}$.
The variance of $\delta_{t+\Delta t}$ is given by

$$Var\{\delta_{t+\Delta t}\} \approx \sum_{i=1}^{n_1}\left(\frac{\partial\{\delta_{t+\Delta t}\}}{\partial a_i}\Bigg|_{a=\bar{a}}\right)^2 \cdot \sigma_i^2 \tag{35}$$

The partial derivative of $\ddot{\delta}_{t+\Delta t}$ with respect to a_i is given by

$$\frac{\partial\{\ddot{\delta}_{t+\Delta t}\}}{\partial a_i} = b_0\left(\frac{\partial\{\delta_{t+\Delta t}\}}{\partial a_i} - \frac{\partial\{\delta_t\}}{\partial a_i}\right) - b_2\frac{\partial\{\dot{\delta}_t\}}{\partial a_i} - b_3\frac{\partial\{\ddot{\delta}_t\}}{\partial a_i} \tag{36}$$

The partial derivative of $\dot{\delta}_{t+\Delta t}$ with respect to a_i is given by

$$\frac{\partial\{\dot{\delta}_{t+\Delta t}\}}{\partial a_i} = \frac{\partial\{\dot{\delta}_t\}}{\partial a_i} + b_6\frac{\partial\{\ddot{\delta}_t\}}{\partial a_i} + b_7\frac{\partial\{\ddot{\delta}_{t+\Delta t}\}}{\partial a_i} \tag{37}$$

The partial derivative of Eq.36 with respect to a_i is given by

$$\frac{\partial^2\{\ddot{\delta}_{t+\Delta t}\}}{\partial a_i^2} = b_0\left(\frac{\partial^2\{\delta_{t+\Delta t}\}}{\partial a_i^2} - \frac{\partial^2\{\delta_t\}}{\partial a_i^2}\right) - b_2\frac{\partial^2\{\dot{\delta}_t\}}{\partial a_i^2} - b_3\frac{\partial^2\{\ddot{\delta}_t\}}{\partial a_i^2} \tag{38}$$

The partial derivative of Eq.37 with respect to a_i is given by

$$\frac{\partial^2\{\dot{\delta}_{t+\Delta t}\}}{\partial a_i^2} = \frac{\partial^2\{\dot{\delta}_t\}}{\partial a_i^2} + b_6\frac{\partial^2\{\ddot{\delta}_t\}}{\partial a_i^2} + b_7\frac{\partial^2\{\ddot{\delta}_{t+\Delta t}\}}{\partial a_i^2} \tag{39}$$

The mean value and variance of the displacement are obtained at time $t+i_1\Delta t(i_1 = 2,3,\cdots,n_3)$ step-by-step.
The partial derivative of Eq.27 with respect to a_i is given by

$$\frac{\partial\{\sigma\}}{\partial a_i} = \frac{\partial[D]}{\partial a_i}[B]\{\delta_{t'}{}^d\} + [D]\frac{\partial[B]}{\partial a_i}\{\delta_{t'}{}^d\} + [D][B]\frac{\partial\{\delta_{t'}{}^d\}}{\partial a_i} \tag{40}$$

The partial derivative of Eq.40 with respect to a_i is given by

$$\frac{\partial^2\{\sigma\}}{\partial a_i^2} = \frac{\partial^2[D]}{\partial a_i^2}[B]\{\delta_{t'}{}^d\} + 2\frac{\partial[D]}{\partial a_i}\frac{\partial[B]}{\partial a_i}\{\delta_{t'}{}^d\} + 2\frac{\partial[D]}{\partial a_i}[B]\frac{\partial\{\delta_{t'}{}^d\}}{\partial a_i}$$

$$+[D]\frac{\partial^2[B]}{\partial a_i^2}\{\delta_{t'}{}^d\} + 2[D]\frac{\partial[B]}{\partial a_i}\frac{\partial\{\delta_{t'}{}^d\}}{\partial a_i}$$

$$+[D][B]\frac{\partial^2\{\delta_{t'}{}^d\}}{\partial a_i{}^2} \tag{41}$$

The stress is expanded at mean value point $\bar{a}=(\bar{a}_1,\bar{a}_2,\cdots,\bar{a}_i,\cdots,\bar{a}_{n_1})^T$ by means of a Taylor series. By taking the expectation operator for two sides of the above Eq.27, the mean of stress is obtained as

$$\mu\{\sigma\}\approx\{\sigma\}\big|_{a=\bar{a}}+\frac{1}{2}\sum_{i=1}^{n_1}\frac{\partial^2\{\sigma\}}{\partial a_i{}^2}\bigg|_{a=\bar{a}}\cdot\sigma_i{}^2 \tag{42}$$

where, $\mu\{\sigma\}$ expresses the mean value of σ.
The variance of σ is given by

$$Var\{\sigma\}\approx\sum_{i=1}^{n_1}\left(\frac{\partial\{\sigma\}}{\partial a_i}\bigg|_{a=\bar{a}}\right)^2\cdot\sigma_i{}^2 \tag{43}$$

6. Numerical example

Figure 1 shows a four-bar linkage, or a crank and rocker mechanism. The establishment of differential equation system can be found in literature 10，11，12.The length of bar 1 is 0.075m, the length of bar 2 is 0.176m, the length of bar 3 is 0.29m,and the length of the bar 4 is0. 286m, the diameters of three bars are 0.02m. The torque T is 4Nm, the load F1 is 20sint N. The three bars are made of steel and they are regarded as three elements. Considering the boundary condition, there are 13 unit coordinates. Young's modulus is regarded as a random variable. For numerical calculation, the means of the Young's modulus within the three bars are . 2×10^{11} . N/m^2 and the variances of the Young's modulus are 10^{11} N^2/m^4 .
Figure 2 shows the mean of the displacement at unit coordinate 11. Unit coordinate 11 is the deformation of the upper end of bar 3 in the vertical direction. The DSFEM simulates 1000 samples. The TSFEM produces an error of less than 0.5%. The CG produces an error of less than 0.1%. Figure 3 shows the variance of the displacement at unit coordinate 11. TSFEM produces an error of less than 1.0%. CG produces an error of less than 0.4%.Figure 4 shows the mean of stress at the top of bar 3. The TSFEM produces an error of less than 0.85%.The CG produces an error of less than 0.13%.Figure 5 shows the variance of stress at the top of bar 3. The TSFEM produces an error of less than 1%. The CG produces an error of less than 0.3%.The results obtained by the CG method and the TSFEM are very close to that obtained by the DSFEM. Table 1 indicates the comparison of CPU time when the mechanism has operated for six seconds.
Figure 6 shows a cantilever beam. The length, the width, the height , the Poisson's ratio ,the Young's modulus and the load F are assumed to be random variables. Their means are 1m, 0.1m, 0.05m, 0.2, 2×10^{11} N/m^2 ,100N.Their standard deviation are 0.2, 0.1, 0.1, 0.01, 10^9, 0.1. Load subjected to the cantilever beam is Fsin(100t)N. It is divided into 400 rectangle elements that have 505 nodes. Figure 7 shows the mean of vertical displacement at node 505. DSFEM simulates 100 samples. The result obtained by the TSFEM produces an error of less than 2% . CG produces an error of less than 0.5%. Figure 8 shows the variance of vertical

displacement at node 505.The TSFEM produces an error of less than 3.0%. CG produces an error of less than 0.8%.Figure 9 shows the mean of horizontal stress at node 5. The TSFEM produces an error of less than 2.4%. CG produces an error of less than 0.9%. Figure 10 shows the variance of horizontal stress at node 5. The TSFEM produces an error of less than 3.2%. CG produces an error of less than 1.3%. Table 2 indicates the comparison of CPU time when the cantilever beam has operated for six seconds.

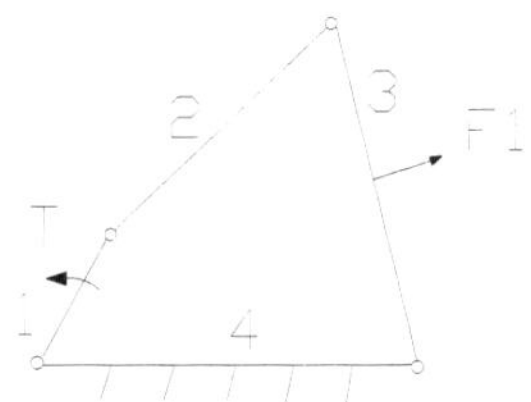

Fig. 1. A four-bar linkage

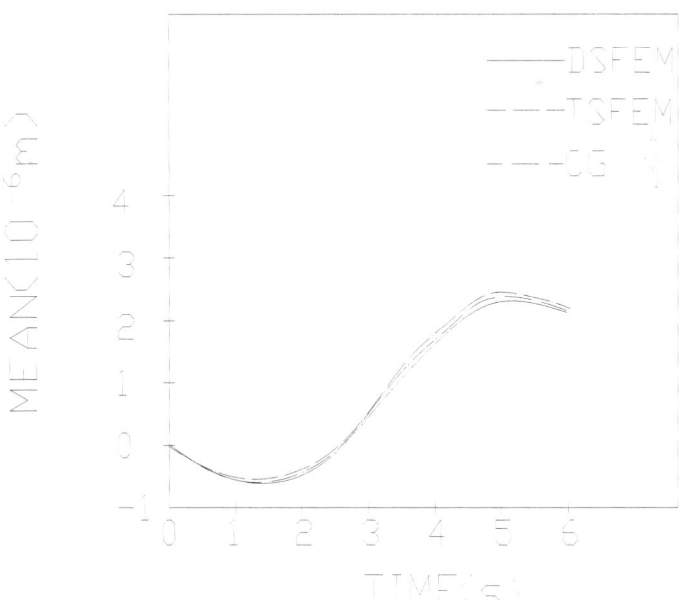

Fig. 2. The mean of displacement at unit coordinate 11 for $\sigma_E{}^2 = 10^{11}$

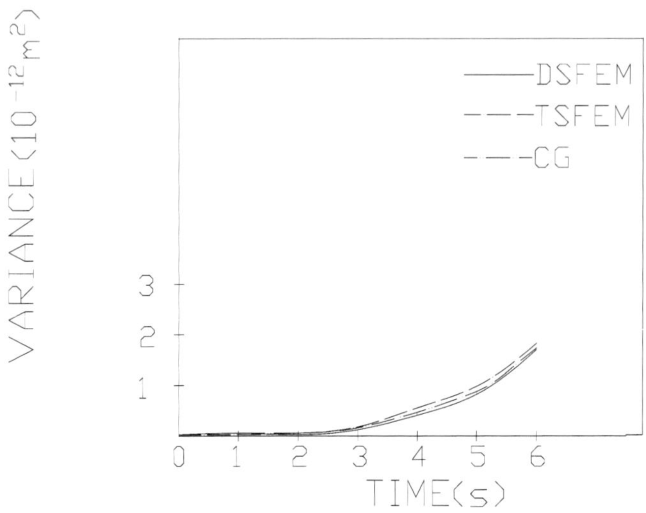

Fig. 3. The variance of displacement at unit coordinate 11 for $\sigma_E{}^2 = 10^{11}$

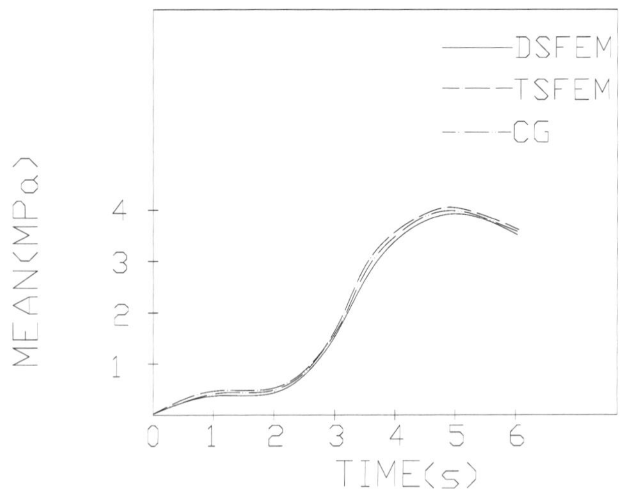

Fig. 4. The mean of stress at the top of bar 3 for $\sigma_E{}^2 = 10^{11}$

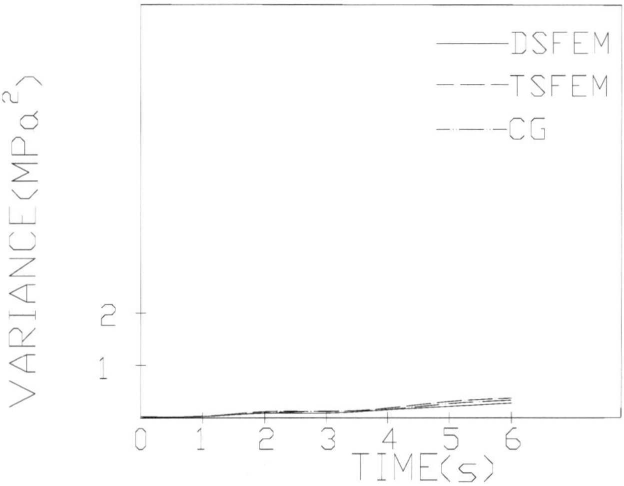

Fig. 5. The variance of stress at the top of bar 3 for $\sigma_E{}^2 = 10^{11}$

DSFEM	TSFEM	CG
19 seconds	4 seconds	14 seconds

Table 1. Comparison of CPU time for $\sigma_E{}^2 = 10^{11}$

Fig. 6. A cantilever beam

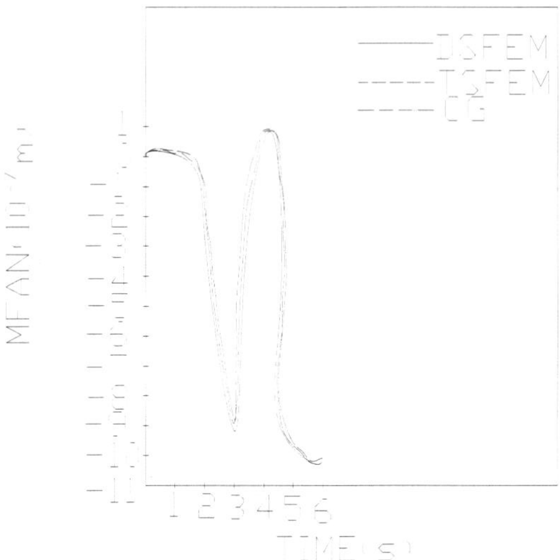

Fig. 7. The mean of vertical displacement at node 505

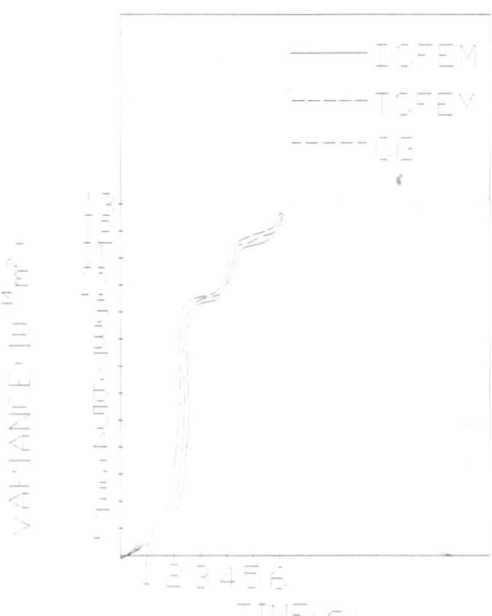

Fig. 8. The variance of vertical displacement at node 505

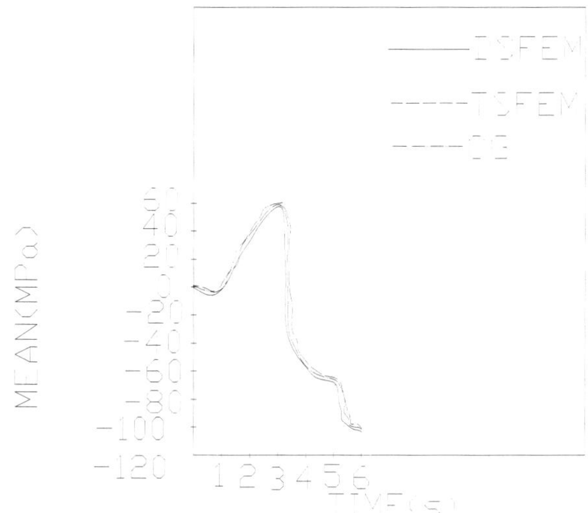

Fig. 9. The mean of horizontal stress at node 5

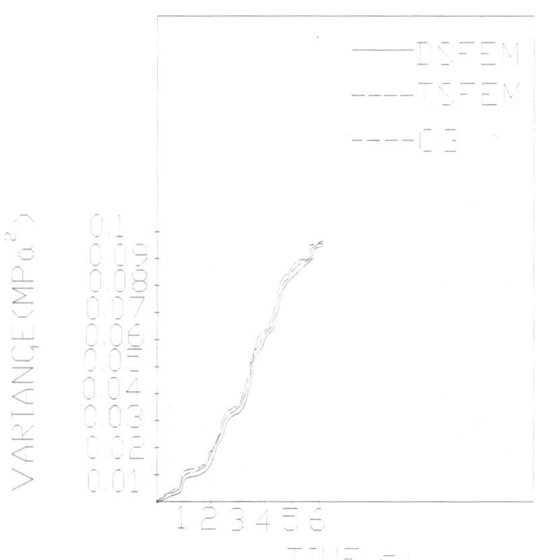

Fig. 10. The variance of horizontal stress at node 5

DSFEM	TSFEM	CG
3 hours 8 minutes 17 seconds	1 hour 45 minutes 10 seconds	40 minutes 24 seconds

Table 2. Comparison of CPU time

7. Conclusions

Considering the influence of random factors, the mechanical vibration in a linear system is presented by using the TSFEM. Different samples of random variables are simulated. The combination of CG method and Monte Carlo method makes it become an effective method for analyzing the vibration problem with the characteristics of high accuracy and quick convergence.

8. References

[1] J. Astill, C. J. Nosseir and M. Shinozuka. Impact loading on structures with random properties.J. Struct. Mech.,1(1972) 63-67

[2] F. Yamazaki , M. Shinozuka and G. Dasgupta.. Neumann expansion for stochastic finite element analysis. J. Engng. Mech. ASCE. 114 (1988):1335-1354

[3] M. Papadrakakis and V. Papadopoulos. Robust and efficient methods for stochastic finite element analysis using Monte Carlo Simulation. Comput. Methods Appl. Mech. Engrg. 134(1996)325 -340

[4] G. B. Baecher and T. S. Ingra. Stochastic FEM in settlement predictions. J. Geotech. Engrg. Div.107 (1981)449-463.

[5] K. Handa and K. Andersson. Application of finite element methods in the statistical analysis of structures. Proc. 3rd Int. Conf. Struct. Safety and Reliability, Trondheim, Norway (1981)409 -417.

[6] T. Hisada and S. Nakagiri. Role of the stochastic finite elenent method in structural safety and reliability. Proc. 4th Int. Conf. Struct. Safety and Reliability ,Kobe, Japan(1985)385-394.

[7] S. Mahadevan and S. Mehta. Dynamic reliability of large frames. Computers & Structures 47 (1993)57-67.

[8] W. K. Liu, T. Belytschko and A. Mani. Probabilistic finite elements for nonlinear structural dynamics. Comput.Methods Appl. Mech. Engrg. 57(1986)61-81.

[9] S.Chakraborty and S.S.Dey. A stochastic finite element dynamic analysis of structures with uncertain parameters. Int. J. Mech. Sci. 40 (1998)1071-1087.

[10] A.G.,Erdman and G.N.,Sandor.A general method for kineto-elastodynamic analysis and synthesis of mechanisms.ASME, Journal of Engineering for Industry 94(1972) 1193-1205.

[11] R.C.Winfrey.Elastic link mechanism dynamics.ASME, Journal of Engineering for Industry 93(1971)268-272.

[12] D.A.,Turcic and A.Midha.Generalized equations of Motion for the dynamic analysis of elastic mechanism system. ASME Journal of Dynamic Systems, Measurement ,and ,Control 106(1984)243-248.

[13] M.Kaminski . Stochastic pertubation approach to engineering structure vibrations by the finite difference method . Journal of Sound and Vibration (2002)251(4), 651—670.

[14] Kaminski,M.On stochastic finite element method for linear elastostatics by the Taylor expansion..Structural and multidisciplinary optimization..35(2008),213-223.

[15] Sachin K;Sachdeva;Prasanth B;Nair;Andy J;Keane.Comparative study of projection schemes for stochastic finite element analysis. Comput. Methods Appl. Mech. Engrg. 195(2006),2371-2392.

[16] Marcin Kaminski.Generalized perturbation-based stochastic finite element in elastostatics. Computer & structures.85 (2007),586-594.